Lecture Notes in Mathematics

Edited by A. Dold and B. Eckmann

820

Willia

The Real Analytic Theory of Teichmüller Space

Springer-Verlag
Berlin Heidelberg New York 1980

Author

William Abikoff
University of Illinois at Urbana-Champaign
Urbana, IL 61801
USA

AMS Subject Classifications (1980): 14H15, 32G15

ISBN 3-540-10237-X Springer-Verlag Berlin Heidelberg New York
ISBN 0-387-10237-X Springer-Verlag New York Heidelberg Berlin

Printed in Germany

Printing and binding: Beltz Offsetdruck, Hemsbach/Bergstr.
2141/3140-543210

CONTENTS

INTRODUCTION

These notes are based on a series of four lectures delivered by Lipman Bers and the author in the "Seminaire sur les diffeomorphismes des surfaces d'apres Thurston" conducted at L'Universite de Paris-sud at Orsay in 1976-77. We attempted to develop some of the classical results and basic machinery in the real analytic theory of Teichmüller space. Bers also gave his relatively recent proof of Thurston's classification of the diffeomorphisms of surfaces.

Little presented here is new, although, to the knowledge of the author, much of the material contained herein does not exist in print. Those results with two major exceptions are folklore in Riemann surface theory. The exceptions are Lemma 3.3.3 of Chapter II, which is an unpublished result due to Bers, and §4 of Chapter III.

Almost all of the contents of these notes is background to or a modest exposition of Bers' work on the problem of moduli.

These notes are complementary to the notes on Thurston's work edited by A. Fathi, F. Laudenbach and V. Poenaru which arose from the same seminar.

I have attempted to keep the presentation at as elementary a level as possible. A general familiarity with the language and notation of mathematics and the definition of a Riemann surface are assumed without apology. The results used without proof are:

1) The uniformization theorem
2) The Riemann-Roch theorem
3) Brouwer's invariance of domain

4) The Ahlfors-Bers theorem giving the solution to the Beltrami equation with dependence on parameters
5) The Hurwitz theorem on the finiteness of the conformal group of a compact surface of genus greater than one
6) Homeomorphic compact Riemann surfaces are diffeomorphic.

References are given at the end of each Chapter. The theorems are referred to as Theorem x.y.z in Chapter w. This result is Theorem z in Section y of §x. Within the same Chapter, the Chapter reference is omitted.

To understand a concept in Riemann surface theory often requires nothing more than to see the relevant picture. The figures accompanying and augmenting the text are but simple examples of the mathematical and artistic skills of George Francis. I expect that the gratitude of the author is inconsequential compared to the gratitude of the reader. What little understanding of the vagaries of the English language which is demonstrated herein must be attributed to the editorial ability and infinite patience of my wife Chris.

The hospitality of L'Institut des Hautes Etudes Scientifiques and the support of the National Science Foundation and the Alfred P. Sloan Foundation greatly facilitated the preparation of this manuscript.

CHAPTER I

THE CLASSICAL RESULTS OF FRICKE AND TEICHMÜLLER ON RIEMANN'S PROBLEM OF THE MODULI OF RIEMANN SURFACES

Let R_g be the set of conformal equivalence classes of Riemann surfaces of genus $g > 1$. Riemann stated without proof that R_g may be holomorphically parametrized by $3g - 3$ complex parameters. It was realized quite soon that R_g has singularities at those surfaces S which admit conformal automorphisms. In approximately 1960, R_g was shown to be a complex analytic variety of dimension $3g - 3$. Almost all proofs of this result depend on the foundational work of Fricke and Teichmüller. Both achieve a parametrization of a ramified covering space of R_g which is real analytic but not *a priori* complex analytic. The covering spaces are cells of real dimension $6g - 6$ which are real analytically equivalent.

Fricke (1897), using the uniformization theorem of Koebe and Poincaré (1907), parametrized the groups of deck transformations covering marked surfaces S of genus g.

Teichmüller studied a set of Riemannian metrics on S with isolated singularities. The space defined by Teichmüller is an open cell admitting a compactification by measured foliations. These foliations have been studied by Hubbard and Masur and Kerckhoff.

We will obtain the results of Fricke and Teichmüller but will avoid the analytic machinery necessary to study quasiconformal mappings with distributional derivatives.

All diffeomorphisms that we consider here are orientation preserving and all surfaces are compact of genus $g > 1$, unless

otherwise stated.

The proofs given here are essentially those of Bers [4].

§1 Uniformization and Fuchsian groups

(1.1) The Uniformization Theorem

It is the opinion of many experts that Riemann surface theory as a subject distinct from the study of complex manifolds, consists of two theorems and their consequences. The second is the subject of this Chapter, the first is the Uniformization Theorem whose statement follows.

Theorem: *If* S *is a Riemann surface, then* S *is holomorphically equivalent to either*

(i) $\mathbb{C}$,

(ii) $\mathbb{C}\setminus\{0\}$,

(iii) $\mathbb{C}/L$ *where* L *is a lattice*,

(iv) $\hat{\mathbb{C}} = \mathbb{C}P^1 = \mathbb{C} \cup \{\infty\}$

(v) U/G *where* $U = \{\operatorname{Im} z > 0\}$ *and* G *is a properly discontinuous group of holomorphic self-maps of* U.

Proof: See Ahlfors [3].

As a first application we classify the holomorphic universal covers of compact Riemann surfaces. If S has genus 0, S is simply connected hence is holomorphically equivalent to $\hat{\mathbb{C}}$. If S has genus 1, then $\pi_1 S$ is a free abelian group of rank 2, and up to holomorphic equivalence, $S = \tilde{S}/G$ where G is the group of cover transformations for the universal cover map $\pi : \tilde{S} \to S$. Since the covering is holomorphic, $G < \operatorname{Aut} \tilde{S}$, where $\operatorname{Aut} \tilde{S}$ is the group of holomorphic automorphisms of $\tilde{S}$. $\tilde{S}$ cannot be holomorphically equivalent to $\hat{\mathbb{C}}$ because the holomorphic automorphism group of $\hat{\mathbb{C}}$ is precisely the group of Möbius transformations

$$\gamma : z \mapsto (az + b)/(cz + d) \quad .$$

Each Möbius transformation has at least one fixed point in $\hat{\mathbb{C}}$, hence the group G does not act properly discontinuously on $\hat{\mathbb{C}}$ unless G is trivial.

The holomorphic automorphism group of U, which we denote $\Gamma = \text{Aut } U$ is a group of Möbius transformations with real entries and is isomorphic to $PSL(2,\mathbb{R})$. The discontinuous subgroups of Γ are the orientation preserving crystallographic groups for the hyperbolic (or Poincaré) plane and are called Fuchsian groups. Real Möbius transformations are classified as follows: if $\gamma : z \mapsto (az + b)/(cz + d)$ with $ad - bc = 1$, then γ is elliptic (respectively, parabolic or hyperbolic) if $|a + d| < 2$ (respectively, $|a + d| = 2$ or $|a + d| > 2$). A parabolic transformation is conjugate in Γ to the translation $z \mapsto z + 1$, hyperbolic γ's are conjugate to $z \mapsto \lambda z$ for some $\lambda \in \mathbb{R}^+\backslash\{1\}$. Elliptic transformations have fixed points in U; thus they do not appear in groups of cover transformations for unramified covers.

We assume $\tilde{S}$ is holomorphically equivalent to U. By the classification given by the uniformization theorem, this hypothesis is valid for all surfaces of genus $g > 1$. The Poincaré (or hyperbolic) metric on U is $ds^2 = y^{-2}(dx^2 + dy^2)$. ds^2 is Γ-invariant, hence projects to a smooth metric on U/G for any properly discontinuous group $G < \Gamma$. Let $\pi : U \to U/G$ denote the projection map.

Lemma 1: If U/G is compact, then each $\gamma \in G\setminus\{id\}$ is hyperbolic.

Proof: Assume $\gamma \in G$ is parabolic. By conjugating G we may assume $\gamma : z \mapsto z + 1$. γ covers a free homotopy class $[\alpha]$ of closed curves on $S = U/G$. By compactness of S

$$\ell[\alpha] = \inf\{\ell(\alpha) | \alpha \in [\alpha]\} > 0$$

where $\ell(\alpha)$ is the Poincaré length of α. For $y_0 > 0$,

$$\alpha_{y_0} = \pi(\{x + iy_0 | x \in [0,1]\}) \in [\alpha]$$

and $\ell(\alpha_{y_0}) = y_0^{-2} \to 0$ as $y_0 \to \infty$ which is a contradiction. ///

Thus we may assume that every non-trivial element of G has two fixed points in $\hat{\mathbb{R}} = \mathbb{R} \cup \{\infty\}$. If G is discontinuous, it is a discrete subgroup of Γ, hence is countable.

Lemma 2: Let $\gamma, \eta \in \operatorname{Aut} U$ be hyperbolic, then γ and η commute if and only if they have the same fixed points.

Proof: Commuting is a conjugation invariant, so we may assume $\gamma : z \mapsto \lambda z$ with $\lambda \neq 1$ and $\eta : z \mapsto (az + b)/(cz + d)$ with $ad - bc = 1$. The composition of Möbius transformations is precisely matrix multiplication. Thus if γ and η commute we have

$$\begin{bmatrix} \lambda^{-1/2} & 0 \\ 0 & \lambda^{1/2} \end{bmatrix} \begin{bmatrix} a & b \\ c & d \end{bmatrix} \begin{bmatrix} \lambda^{1/2} & 0 \\ 0 & \lambda^{-1/2} \end{bmatrix} \begin{bmatrix} d & -b \\ -c & a \end{bmatrix} = \begin{bmatrix} 1 & 0 \\ 0 & 1 \end{bmatrix}$$

A short computation shows that $b = c = 0$, i.e., $\eta : z \mapsto \lambda_1 z$. ///

An immediate consequence of the Lemma is that $\operatorname{Aut} U$ has no free abelian discontinuous subgroups of rank 2 with hyperbolic generators. Thus a surface S of genus 1 has $\tilde{S} \simeq \mathbb{C}$. A Riemann surface S is called <u>hyperbolic</u> (respectively, <u>parabolic</u> or <u>elliptic</u>) if $\tilde{S}$ is holomorphically equivalent to U (respectively, $\mathbb{C}$ or $\hat{\mathbb{C}}$). We then have

<u>Proposition 1</u>: If g denotes the genus of the compact Riemann surface S, then $g = 0$ (respectively, $g = 1$ or $g > 1$) whenever S is elliptic (respectively, parabolic or hyperbolic).

<u>Proposition 2</u>: If S has genus > 1 and $\varphi : S \to S$ is a holomorphic automorphism homotopic to the identity then $\varphi = \text{id}$.

<u>Proof</u>: Represent S as U/G and let $\varphi_t : S \to S$ be the homotopy of φ to the identity. φ is covered by a Möbius transformation γ which normalizes G. $\varphi = \text{id}$ if and only if $\gamma = \text{id}$. If $\tilde{\alpha}$ denotes the Poincaré geodesic in the free homotopy class of α, then $\tilde{\alpha} = \widetilde{\varphi_t(\alpha)}$. $\tilde{\alpha}$ is covered by some $\eta \in G$ whose fixed points are precisely the endpoints of a component of $\pi^{-1}(\tilde{\alpha})$. Thus γ fixes these endpoints. Since $\pi_1 S$ is not cyclic, γ fixes more than two points, hence $\gamma = \text{id}$. ///

(1.2) <u>Marked Riemann Surfaces and Fricke Space</u>

Let S be a compact Riemann surface of genus g. A <u>marking</u> C of S is an ordered set of standard generators $\alpha_1, \ldots, \alpha_{2g}$ for $\pi_1 S$, i.e., a set of simple loops satisfying:

(1) $\displaystyle\prod_{i \text{ odd}} [\alpha_i, \alpha_{i+1}] = 1$ where $[\ ]$ denotes the commutator,

and

(2) the geometric intersection number $\#$ is, for $i \leq j$,

$$\#(\alpha_i, \alpha_j) = \begin{cases} \delta_{i+1,j} & \text{for } i \text{ odd} \\ 0 & \text{for } i \text{ even} \end{cases} .$$

The pair (S,C) is called a marked surface.

We next consider the set Ξ of all monomorphisms $\chi : \pi_1 S \to \Gamma = \mathrm{Aut}\, U$ with the following properties:

(1) $\chi(\pi_1 S)$ is discrete

(2) for a fixed marking C of S

$\chi(\alpha_{2g})$ has 0 as a repelling fixed point

$\chi(\alpha_{2g})$ has ∞ as an attractive fixed point

$\chi(\alpha_{2g-1})$ has 1 as a fixed point.

Lemma: Let

$$\gamma_i = \chi(\alpha_i) : z \mapsto (a_i z + b_i)/(c_i z + d_i)$$

with $a_i d_i - b_i c_i = 1$ and $a_i \geq 0$ and $b_i > 0$ if $a_i = 0$. Then the map

$$\mathfrak{Z}_g : \Xi \to R^{6g-6}$$

$$\chi \mapsto (a_1, b_1, c_1, \ldots, a_{2g-2}, b_{2g-2}, c_{2g-2})$$

is injective.

Proof: The commutator relation $\prod\limits_{i\ \mathrm{odd}} [\alpha_i, \alpha_{i+1}] = 1$ gives the matrix equation

$$\eta\, \gamma_{2g-1} = \gamma_{2g}\, \gamma_{2g-1}\, \gamma_{2g}^{-1}$$

where

$$\eta = \prod_{\substack{i\ \mathrm{odd} \\ i < 2g-2}} [\gamma_i, \gamma_{i+1}] : z \mapsto (a'z + b')/(c'z + d') \quad .$$

We must show that the above matrix equation, together with the knowledge of $\mathfrak{Z}_g(\chi)$ allows us to solve for the coefficients of γ_{2g-1}

and γ_{2g}. By definition of χ, $c_{2g} = b_{2g} = 0$ and we write $\lambda = a_{2g} = d_{2g}^{-1}$. The matrix equation, together with the relation $a'd' - b'c' = 1$, implies that the coefficients satisfy

[1] $$(a'-1)a_{2g-1} + b'c_{2g-1} = 0$$

[2] $$c'a_{2g-1} + (d'-\lambda^{-2})c_{2g-1} = 0$$

[3] $$c'b_{2g-1} + (d'-1)d_{2g-1} = 0$$

[4] $$(a'-\lambda^2)b_{2g-1} + b'd_{2g-1} = 0 \quad .$$

Since $\gamma_{2g-1}(1) = 1$ we also have

[5] $$a_{2g-1} + b_{2g-1} = c_{2g-1} + d_{2g-1} \quad .$$

Equations [1] - [4] have the unique solution

$$a_{2g-1} = b_{2g-1} = c_{2g-1} = d_{2g-1} = 0$$

unless $\lambda^2 = (1-a')/(d'-1)$. Since the trivial solution is not possible, we may take λ to be the positive square root of $(1-a')/(d'-1)$. Equations [1], [3], [5] and the determinant condition

$$a_{2g-1}d_{2g-1} - b_{2g-1}c_{2g-1} = 1$$

then have a unique solution with $a_{2g-1} \geq 0$. The rest of the coefficients then follow and the map $\mathfrak{F}_g$ is injective. ///

<u>Definition</u>: $F_g = \mathfrak{F}_g(\Xi)$ is called the <u>Fricke space</u> of genus g (> 1).

<u>Theorem</u>: F_g <u>is a parametrization of the marked Riemann surfaces of genus</u> g.

Proof: If $x \in F_g$ then x determines a Fuchsian group G_x and a standard set of generators $\gamma_1,\ldots,\gamma_{2g}$ of G_x. These generators cover a marking $\alpha_1,\ldots,\alpha_{2g}$ of the surface U/G_x. Thus x determines a marked surface. The converse is a direct, but highly non-trivial consequence of the Uniformization Theorem. ///

§2 Grötzsch's Problem

In the late 1920's, Grötzsch posed and solved an elementary extremal problem in conformal mapping. The extremal map is affine. The same extremal problem may be posed for compact surfaces of genus greater than one. The solution to the latter problem is the Teichmüller theorems which are the subject of this Chapter. Teichmüller's results are far more complicated to prove for two reasons. The first is that on a manifold a map will be affine only in terms of fixed choices of local coordinate, i.e., affinity is not intrinsic. The second is that the Grötzsch problem has boundary data, while on compact surfaces there is no boundary. Teichmüller resolves the first difficulty using quadratic differentials and the second using an ergodic argument.

(2.1) The Grötzsch problem

Let R and R' be rectangles in the complex plane. For purposes of normalization, we assume that the ordered vertices of R are $V = \{0,a,a+ib,ib\}$ and of R' are $V' = \{0,a',a'+ib',ib'\}$ with $a,b,a',b' > 0$. For a $\mathcal{C}^1$-diffeomorphism $f\colon R \to R'$ let

$$K_f(z) = \frac{|f_z(z)| + |f_{\bar z}(z)|}{|f_z(z)| - |f_{\bar z}(z)|}$$

and

$$K[f] = \|K_f(z)\|_\infty .$$

<u>Theorem</u>: (Grötzsch) <u>If</u> $f:R \to R'$ <u>is a</u> $\mathcal{C}^1$<u>-diffeomorphism, mapping</u> V <u>to</u> V' <u>and preserving the ordering, then</u> $K[f] \geq K_0 = (a'/a)/(b'/b)$ <u>with equality if and only if, for</u> $z = x + iy$,

$f(z) = f_0(z) = (a'/a)x + i(b'/b)y.$

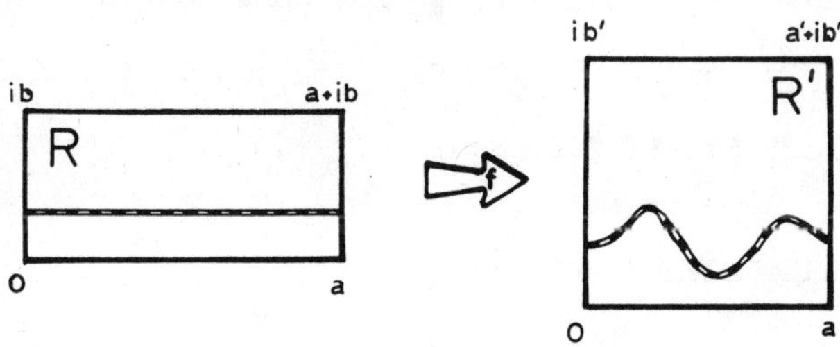

Figure 1

<u>Proof</u>: Let α be a horizontal curve as shown in Figure 1. Then

$$a' \leq \int_a |f_x| \, dx .$$

Integrating with respect to y, we obtain

$$a'b \leq \iint_R |f_x| \, dxdy.$$

Since $f_x = f_z + f_{\bar{z}}$ and the jacobian of f is $j_f(z) = |f_z|^2 - |f_{\bar{z}}|^2$,

$$a'b \leq \iint_R (|f_z| + |f_{\bar{z}}|) \, dxdy$$

$$= \iint_R K_f(z)^{1/2} j_f(z)^{1/2} dxdy \quad .$$

We then apply the Schwarz inequality and square to obtain

$$a'b \leq K[f] ab'$$

which is the desired inequality.

We examine the conditions for equality to hold. First set $u + iv = f(x+iy)$. From Equation [1] we see that $v_x = 0$, for $y = y_0$ fixed. Doing the identical computation on vertical lines gives $u_y = 0$ for $x = x_0$ fixed. Thus $f(x+iy) = u(x) + iv(y)$ and $j_f = u_x v_y$. Since $j_f \neq 0$, $u_x > 0$ and $v_y > 0$. Further $K_f = (u_x)/(v_y)$. For equality in the application of the Schwarz inequality, we must have $K_f = cj_f = K[f]$ for some constant c. It follows that $v = c_1 y$ and $u = c_2 x$ for some constants c_1 and c_2. The constants are trivially as required. ///

§3 Teichmüller's Uniqueness Theorem

(3.1) The geometry of quadratic differentials

Let S be a Riemann surface. A quadratic differential on S is an expression $\omega = \varphi(z)dz^2$ in terms of a local coordinate z on S, which is invariant under change of local coordinate. This means that if z_1 is another local coordinate on S and $\omega = \varphi_1(z_1)dz_1^2$ then $\varphi_1(z_1)\left[\frac{dz_1}{dz}\right]^2 = \varphi(z)$. ω is holomorphic if φ is. We restrict out attention to the case where S is compact of genus $g > 1$ and $\omega \in \mathcal{Q}_S$, the holomorphic quadratic differentials on S. In a language which we will not use, $\mathcal{Q}_S = H^0(S, \Omega_S^{\otimes 2})$ where Ω_S is the canonical bundle on S.

Notice that the zeroes of ω and the orders of those zeroes do not depend on the choice of local coordinate whenever $\omega \in \mathcal{Q}_S \backslash \{0\}$. We denote by b_ω the set of zeroes of ω.

For $p \in S$ and $\omega \in \mathcal{Q}_S \backslash \{0\}$ with $\omega(p) \neq 0$, suppose z is a local coordinate at p with $z(p) = 0$. Then

$$\zeta(z) = \int_0^z \sqrt{\omega} = \int_0^z \varphi^{1/2} dz$$

is a local coordinate, i.e., holomorphic and injective, near p for some fixed choice of sign for the square root. ζ is called the ω-<u>coordinate</u> at p. $\zeta^{-1}(\mathbb{R})$ (respectively, $\zeta^{-1}(i\,\mathbb{R})$) is called the <u>horizontal</u> (respectively, <u>vertical</u>) <u>line</u> through p. These lines are completely characterized by being the unique analytic curves through p on which $\omega > 0$ (respectively, $\omega < 0$). Since this construction is independent of the choice of local coordinate we immediately obtain two transverse foliations of $S \backslash b_\omega$.

We examine the structure of the singularities of these foliations. If ω has a zero of order m at $p \in S$, then in some local coordinate near p,

$$\omega = z_1^m dz_1^2 \quad .$$

If m is even, $\sqrt{\omega}$ has a single valued branch and ζ maps a neighborhood of p to a neighborhood of 0 and is $(m+2)/2$-sheeted and ramified over 0. Thus $\zeta^{-1}(\mathbb{R})$ consists of $m + 2$ analytic rays, which are equally spaced, emanating from p. If m is odd, set $z_2^2 = z_1$, i.e., form a two-sheeted cover R of a neighborhood of p. Then

$$\omega(z_1) = \omega_2(z_2) = 4\, z_2^{2m+2} dz_2^2$$

is locally a quadratic differential on R

$$\zeta_2 = \int_0^{z_2} \sqrt{\omega_2} = (m+2)^{-1} z_2^{m+2}$$

is a local coordinate on R. If $\zeta_2(z_2)$ is real, then $\omega_2(z_2) \geq 0$ and $\omega(z_1) \geq 0$ and conversely. $\zeta_2^{-1}(\mathbb{R})$ consists of $m + 2$ analytic curves through $z_2^{-1}(0)$. When this is projected to S we obtain $m + 2$ equally spaced analytic rays emanating from p.

The same argument shows that the foliation of S by vertical lines has singularities of the same form at the points where $\omega = 0$. Notice that the foliations are invariant under multiplication of ω by a positive real constant.

The foliations defined above are called the ω-horizontal and ω-vertical foliations of S. The leaves of these foliations have no global orientation since it is generally impossible to choose a globally defined sign for $\sqrt{\omega}$. This will cause some difficulty in the proof of Lemma 2.5. We let S^o be the two-sheeted cover of S to which $\sqrt{\omega}$ lifts as a single-valued differential. S^o is called the orientation cover of S. Leaves lift to leaves but on S^o they may be given a consistent orientation.

(3.2) Measures on the horizontal and vertical foliations

For each $\omega \in \mathcal{Q}_S \backslash \{0\}$, we may obtain transverse measures for the horizontal and vertical foliations quite simply. The two cases are symmetric; here we only consider the ω-horizontal foliation. If α is a piecewise-differentiable arc lying in a simply connected subset of $S \backslash b_\omega$ set $\mu(\alpha) = V[\mathrm{Im}\, \zeta(\alpha)]$ where V denotes total variation and ζ is as in the previous section. This trivially extends to a measure on all piecewise-differentiable curves, since the singularities of ω are of no importance.

(3.3) Differential geometry of quadratic differentials

Let $\omega \in \mathcal{Q}_S \setminus \{0\}$. We define invariant area and line elements of S using the quadratic differential ω.

Definition: If $p \in S$ is not a zero of ω and z is a local coordinate at p, then $\omega = \varphi(z)dz^2$.

$$dA_\omega = |\varphi| \, dxdy = (2i)^{-1} |\varphi| \, dz \wedge d\bar{z}$$

is the ω-area element on S.

$$ds_\omega = |\varphi|^{1/2} |dz|$$

is the ω-line element on S.

Notice that the zeroes of ω are irrelevant and $A_\omega(S) = \iint_S dA_\omega$ is finite. Further ds_ω is a Euclidean line element in terms of the ω-coordinate on S if we avoid the singularities of ω; i.e., if $\zeta = \xi + i\eta$ is the ω-coordinate, $ds_\omega^2 = d\xi^2 + d\eta^2$. The length of a piecewise differentiable arc $\alpha \subset S$ is $\ell_\omega(\alpha) = \int_\alpha ds_\omega$. Next we propose to show that there is a unique curve in each homotopy class of curves from p_0 to p_1 which minimizes ℓ_ω. By abuse of language we call this curve the (ω-) geodesic from p_0 to p_1. Since ds_ω is locally Euclidean, a short geodesic (avoiding b_ω) is an arc α along which ζ or ω has constant argument.

Let $\tilde{S}$ be the universal cover of S. $\omega \in \mathcal{Q}_S \setminus \{0\}$ lifts to a quadratic differential $\tilde{\omega}$ on $\tilde{S}$. Using the Uniformization Theorem, $\tilde{S}$ is holomorphically equivalent to $\Delta = \{|z| < 1\}$ and there is a global coordinate z on $\tilde{S}$. Thus there is a holomorphic function $\tilde{\varphi}$ on $\tilde{S}$, which we identify with Δ, so that $\tilde{\omega} = \tilde{\varphi}(z)dz^2$ globally.

<u>Lemma 1</u>: Let P be a polygon in $\tilde{S}$ with $\tilde{\omega}$-geodesic sides. Then, for $m \geq 0$, P has at least 3 vertices with angles $< (m+2)^{-1} 2\pi$ at zeroes of $\tilde{\omega}$ of order m. These angles are measured in the hyperbolic (or ordinary Euclidean) metric on Δ.

<u>Proof</u>: Let n be the number of zeroes of $\tilde{\varphi}$ inside P counted according to multiplicity. Further, let θ_i be the angle at the vertex v_i and m_i the order of the zero of $\tilde{\varphi}$ at v_i. A standard residue computation similar to the argument principle shows that

$$\int_{\partial P} d(\arg \tilde{\varphi}) = 2n\pi - \Sigma\, m_i \theta_i \quad .$$

Since each side of P is geodesic, on each side

$$0 = \arg d\tilde{\zeta} = (1/2) \arg \tilde{\varphi} + \arg dz$$

where $\tilde{\zeta}$ is the $\tilde{\omega}$-coordinate on Δ. Also

$$\int_{\partial P} d(\arg dz) + \Sigma(\pi - \theta_i) = 2\pi \quad .$$

The last three equations give

$$\sum_i \{2\pi - (m_i + 2)\theta_i\} = (n+2)2\pi \quad .$$

Each term on the right hand side is dominated strictly by 2π; the left hand side is $\geq 4\pi$. Thus at least three terms in the sum are positive, which is the required result. ///

<u>Theorem</u>: <u>On a compact Riemann surface</u> S <u>of genus greater than</u> 1, <u>each homotopy class with fixed endpoints contains a unique</u> ω<u>-geodesic</u>.

Proof: If there are two geodesics between p and q, then there is a polygon P with geodesic sides whose boundary consists of the shortest curves between two vertices p' and g'. P is simply connected, so it lifts to a polygon $\tilde{P}$ in $\tilde{S}$. If a vertex v_i of $\tilde{P}$ does not cover p' or q', then the angle θ_i at v_i satisfies $\theta_i \geq (m_i+2)^{-1}2\pi$, since the arcs $\{\arg \zeta = \text{constant}\}$ are spaced at angles $(m_i+2)^{-1}2\pi$ at a vertex v_i of order m_i. This contradicts the previous Lemma. Existence of ω-geodesics in each homotopy class with fixed endpoints follows directly from the compactness of S. ///

Lemma 2: Let $f:S \to S$ be homotopic to the identity and $\omega \in \mathcal{Q}_S\backslash\{0\}$. Then there exists a constant $M > 0$ so that for all ω-horizontal arcs α not passing through a zero of ω

$$\ell_\omega(f(\alpha)) \geq \ell_\omega(\alpha) - 2M \quad .$$

Proof: Let $F:S \times I \to S$ be the given homotopy and

$$\alpha_p = f_t(p) = \{F(p,t) \mid t \in I\} \quad .$$

Further let $\bar{\alpha}_p$ be the geodesic in the homotopy class of α_p (with fixed endpoints) and $M = \sup\{\ell_\omega(\bar{\alpha}_p) \mid p \in S\}$. $M < \infty$ since $\ell_\omega(\bar{\alpha}_p)$ is a continuous function on S and S is compact. If α goes from p to q, then $\alpha_p f(\alpha)\alpha_q^{-1}$ is homotopic to α with fixed endpoints. The same is true for $\bar{\alpha}_p f(\alpha)\bar{\alpha}_q^{-1}$. Thus

$$\ell_\omega(\alpha) \leq \ell_\omega(\bar{\alpha}_p) + \ell_\omega(f(\alpha)) + \ell_\omega(\bar{\alpha}_q)$$

$$\leq \ell_\omega(f(\alpha)) + 2M$$

since α is an ω-geodesic arc. ///

(3.4) Teichmüller deformations

Let $\omega \in \mathcal{Q}_S \backslash \{0\}$ and $k \in [0,1)$. We define a deformation of the complex structure of S as follows. Let $p_0 \in S \backslash b_\omega$ and ζ be the ω-coordinate at p_0 with $\zeta(p_0) = 0$. A continuous germ f_{p_0} at p_0 is holomorphic in the $S' = (S,\omega,k)$ structure on S if $f_{p_0} = f_1 \circ \zeta'$ where $\zeta' = (\zeta+k\bar{\zeta})/(1-k)$ and f_1 is some holomorphic germ at $0 \in \mathbb{C}$. If $k = 0$, $f_{p_0} \in \mathcal{O}_{p_0}(S)$, the ring of germs of functions at p_0 which are holomorphic in the original structure on S. In terms of the ω'-coordinate ζ' on S', $\zeta = \xi + i\eta$ and $\zeta' = \xi' + i\eta'$, we have

$$\eta' = \eta$$

$$\xi' = K\xi \quad \text{where} \quad K = (1+k)/(1-k) \quad .$$

$\omega' = (d\zeta')^2 \in \mathcal{Q}_{S'} \backslash \{0\}$ is the quadratic differential on S' induced by the deformation (S,ω,k). The transformation from the ζ coordinate on S to the ζ' coordinate on S' leaves invariant the length of vertical lines, but stretches horizontal lines by a factor K.

The structure of S' as a Riemann surface has only been defined for $p \in S \backslash b_\omega$. To show that S' is indeed a Riemann surface, we must show that a small punctured disc D at $p \in b_\omega$ is conformally a punctured disc on S', not an annulus. Assume p is a zero of even order m of ω, i.e., $\omega = z^m dz^2$ for appropriate choice of local coordinate z. Set $z_1 = z^{(m+2)/2}$, i.e., choose a local $(m+2)/2$ sheeted covering of D. ω lifts to the quadratic differential $\tilde{\omega} = 4(m+2)^{-2} dz_1^2$. A horizontal (respectively, vertical) line of $\tilde{\omega}$ is the lift of a horizontal (respectively, vertical) line of ω. Let ζ_1 be the $\tilde{\omega}$-coordinate at a lift of p. The ω-coordinate near p lifts to the $\tilde{\omega}$-coordinate $\zeta_1' = (\zeta_1 + k\bar{\zeta}_1)/(1-k)$. Projecting to D we obtain

$$[1] \qquad \zeta' = (\zeta_1')^{(m+2)/2} = \left\{\frac{2}{m+2} \quad \frac{\zeta^{(m+2)/2} + k\bar{\zeta}^{(m+2)/2}}{1-k}\right\}^{2/(m+2)}$$

as the ω'-coordinate at p. If m is odd, we first lift ω to the orientation cover and then repeat the above argument to again obtain [1]. S' thus has the structure of a Riemann surface since we have given a local coordinate at each $p \in S'$.

<u>Definition</u>: The Riemann surface structure $S' = (S,\omega,k)$ is called a <u>Teichmüller deformation</u> of the Riemann surface structure S.

<u>Proposition</u>: Let $S' = (S,\omega,k)$. f is a holomorphic germ at $p_0 \in S\backslash b_\omega$ is and only if $f = f_1 \circ \zeta'$ and f_1 satisfies

$$(f_1)_{\bar{\zeta}} = k(f_1)_{\zeta} \quad .$$

<u>Proof</u>: f is a holomorphic germ on S' precisely when $f = f_1 \circ \zeta'$ and f_1 is locally a holomorphic function of ζ'. An equivalent condition is that $(f_1)_{\overline{\zeta'}} = 0$. The chain rule then gives:

$$0 = \frac{\partial f_1}{\partial \overline{\zeta'}} = \frac{\partial f_1}{\partial \zeta}\frac{\partial \zeta}{\partial \overline{\zeta'}} + \frac{\partial f_1}{\partial \bar{\zeta}}\frac{\partial \bar{\zeta}}{\partial \overline{\zeta'}}$$

$$= -k\frac{\partial f_1}{\partial \zeta} + \frac{\partial f_1}{\partial \bar{\zeta}} \quad . \qquad ///$$

Note that ω and $k_1\omega$ differ in their adapted coordinates by a scalar factor which is clearly irrelevant for deforming conformal structure. It follows that the deformations of complex structure obtained as Teichmüller deformations may be parametrized by those ω lying in the unit sphere Σ_S of $\mathcal{Q}_S$ in any norm and $k \in]0,1[$. The base structure S corresponds to $k = 0$.

Definition: The Teichmüller space based at S is the set

$$T(S) = \{(S,\omega,k) \mid \omega \in \Sigma_S, k \in]0,1[\} \cup \{0\}$$

with the obvious topology.

As $k \to 0$, $(S,\omega,k) \to S$. For $\omega \in \Sigma_S$ fixed, we consider the ray $\{(S,\omega,k) \mid k \in]0,1[\}$. The vertical measured foliation of S changes as a measured foliation (in fact degenerates) as $k \to 1$. It is important to note that the horizontal foliation of (S,ω,k) is independent of k. Thus the space of Teichmüller deformations admits a compactification by measured foliations. We summarize these results in the following

Theorem: Let S be a compact Riemann surface of genus $g > 1$ and $S' = (S,\omega,k)$ for some $\omega \in \Sigma_S$ and $k \in]0,1[$. If $\omega' \in \mathcal{Q}_{S'}$ is induced by ω, then the ω' horizontal foliation of S' is independent of k. In particular, $T(S)$ admits a compactification $\overline{T}(S)$ whose boundary points are the ω' horizontal measured foliations of (S,ω,k).

It is an immediate consequence of the Riemann-Roch theorem that $\mathcal{Q}_S$ has complex dimension $3g - 3$. Proofs of this fact, in the language of automorphic forms, may be found in Lehner's A Short Course in Automorphic Functions or Shimura's Introduction to the Arithmetic Theory of Automorphic Functions. Thus $T(S)$ is the open $6g - 6$ dimensional disc and $\overline{T}(S)$ is the $6g - 6$ dimensional closed disc. The remainder of this Chapter is devoted to relating $T(S)$ to conformal structures on S and giving a base-point-free metric description of $T(S)$.

(3.5) Teichmüller's Uniqueness Theorem

The proof of the Teichmüller uniqueness theorem is an example of length-area arguments which were introduced classically, e.g., to study the modulus of an annulus, and first used extensively by Grötzsch (c.f. §2). An invariant formulation has been given by Ahlfors and Beurling.

Let $\omega \in \mathcal{Q}_S \setminus \{0\}$, $k \in]0,1[$, $S' = (S,\omega,k)$ and ω' the quadratic differential on S' induced by ω. Let f be a diffeomorphism of S possibly with isolated singularities. The singular set of f we denote b_f. If $\zeta \in S\setminus(b_f \cup b_\omega)$, let $\lambda_{f,\omega,\omega'}(\zeta)$ be the infinitesimal stretching of the horizontal line through ζ induced by f considered as a map $S \to S'$ (i.e., from the ω to the ω'-coordinate). In these same coordinates, let $j_{f,\omega,\omega'}(\zeta)$ be the jacobian.

Since the underlying set is S, we may also measure the horizontal stretching and jacobian solely in terms of the ω-coordinate. These will be denoted $\lambda_{f,\omega,\omega}$ and $j_{f,\omega,\omega}$ respectively.

Let $\zeta = \xi + i\eta$ be the ω-coordinate on S. We define the astigmatism (classically the dilatation) of f

[1] $$K_f(\zeta) = \frac{|f_\zeta| + |f_{\bar\zeta}|}{|f_\zeta| - |f_{\bar\zeta}|}$$

where $\zeta' = f(\zeta) = \xi' + i\eta$ is the ω'-coordinate. If $K_f(\zeta) \neq 0, \infty$ we note that

[2] $$K_f(\zeta) + K_f(\zeta)^{-1} \geq [(\xi'_\xi)^2 + (\xi'_\eta)^2 + (\eta'_\xi)^2 + (\eta'_\eta)^2]\, j_{f,\omega,\omega}$$

since

[3] $$j_{f,\omega,\omega'} = (\xi'_\xi)(\eta'_\eta) - (\xi'_\eta)(\eta'_\xi) \quad .$$

<u>Definition</u>: A diffeomorphism (possibly with isolated singularities) $f:S \to S'$ is <u>admissible</u> if

$$\|K_f\|_\infty = K[f] < \infty \quad .$$

<u>Lemma 1</u>: If $f:S \to S$ is an admissible diffeomorphism homotopic to the identity, then

[4] $$A_\omega(S) \leq \iint_S \lambda_{f,\omega,\omega} dA_\omega \quad .$$

(i.e., the mean horizontal stretching is greater than or equal to one.)

<u>Proof</u>: Passing to the orientation cover S^o of S and the differential $\tilde{\omega}$ covering ω precisely doubles both sides of equation [4]. Thus we may assume ω has a holomorphic square root and the ω-horizontal foliation has a global orientation on $S \backslash b_\omega$.

Let α be a horizontal line of length L which avoids b_ω. If p is the midpoint of α, set

$$\delta(p,L) = \ell_\omega(f(\alpha)) = \int_\alpha \lambda_{f,\omega,\omega} ds_\omega \quad .$$

Then

$$\tilde{\delta}_L = \iint_S \delta(p,L)\, dA_\omega(p)$$

$$= \iint_S [\lambda_{f,\omega,\omega} ds_\omega]\, dA_\omega(p)$$

since we may ignore the case where p lies on a horizontal line through $b_f \cup b_\omega$. Starting at any basepoint $p_0 \in S \backslash b_\omega$,

$$\zeta(p) = \int_{p_0}^{p} \omega^{1/2}$$

is defined up to an additive constant. Set $\zeta(p) = \xi(p) + i\eta(p)$. For any integrable function $g: S \to \mathbb{C}$ and any fixed real number τ.

$$\iint_S g(p)\,dA_\omega(p) = \iint_S g(\zeta(p))\,d\xi d\eta$$

$$= \iint_S g(\zeta(p)+\tau)\,d\xi d\eta$$

and

$$\delta(p,L) = \int_{-L/2}^{L/2} \lambda_{f,\omega,\omega}(\xi(p)+\tau+i\eta(p))\,d\tau \quad .$$

Thus

$$\tilde{\delta}_L = \iint_S \left[\int_{-L/2}^{L/2} \lambda_{f,\omega,\omega}(\xi(p)+\tau+i\eta(p))\,d\tau \right] d\xi d\eta$$

$$- \iint_S \left[\int_{-L/2}^{L/2} \lambda_{f,\omega,\omega}(u(p)+i\eta(p))\,d\tau \right] du d\eta$$

where $u = \xi(p) + \tau$. Consequently

[5] $$\tilde{\delta}_L = L \iint_S \lambda_{f,\omega,\omega}\,dA_\omega$$

and, by Lemma (3.3),

[6] $$\ell_\omega(f(\alpha)) \geq L - 2M \quad .$$

By [5] and [6],

$$L\iint_S \lambda_{f,\omega,\omega}\,dA_\omega \geq (L-2M)A_\omega(S) \quad .$$

Now divide by L and let $L \to \infty$ to obtain the desired result. ///

<u>Corollary</u>: If $h:S' \to S'$ is an admissible diffeomorphism homotopic to the identity and $\omega' \in \mathcal{Q}_{S'}\setminus\{0\}$, then

$$A_{\omega'}(S') \leq \iint_{S'} \lambda^2_{h,\omega',\omega'}\,dA_{\omega'} \quad .$$

<u>Proof</u>: By the Schwarz Inequality and the Lemma we have

$$A_{\omega'}(S') \leq \iint_{S'} \lambda_{h,\omega',\omega'}\,dA_{\omega'} \leq \iint_{S'} \lambda^2_{h,\omega',\omega'}\,dA_{\omega'} \quad . \qquad ///$$

<u>Lemma 2</u>: The composition of admissible diffeomorphisms is admissible.

<u>Proof</u>: Let D be open in $\mathbb{C}$ and $F:D \to \mathbb{C}$ be differentiable with $F_z \neq 0$. If $\mu_F(z) = (F_{\bar{z}})/(F_z)$ then

$$\text{[7]} \qquad K_F(z) = \frac{1 + |\mu_F|}{1 - |\mu_F|} \quad .$$

Let $f:S \to S'$ and $g:S' \to S''$ be admissible diffeomorphisms. Since $g \circ f$ has isolated singularities, to show that $g \circ f$ is admissible we need only show that $\| K_{g \circ f} \|_\infty < \infty$. We perform the following computations away from the irrelevant singular set. Let z be a local coordinate at $p \in S$ and $w = f(z)$ a local coordinate at $p' \in S'$. Using any local coordinate at $g \circ f(p)$, g may be considered as a map from a disc D in $\mathbb{C}$ into $\mathbb{C}$. The chain rule then gives

$$(g \circ f)_z = g_w w_z + g_{\bar{w}} \bar{w}_z$$

$$(g \circ f)_{\bar{z}} = g_w w_{\bar{z}} + g_{\bar{w}} \bar{w}_{\bar{z}} \quad .$$

An immediate consequence is that $K_{g \circ f}(p)$ is independent of the choice of local coordinates on S, S' and S''. Also $K[g \circ f] < \infty$ if and only if $\|\mu_{f \circ g}\|_\infty < 1$. Since $\bar{w}_{\bar{z}} = \overline{(w_z)}$ and $\bar{w}_z = \overline{(w_{\bar{z}})}$ we have

[8] $$\mu_{g \circ f}(z) = \frac{\mu_f(z) + \mu_g(w)\exp(-2i \arg w_z)}{1 + \bar{\mu}_f(z)\mu_g(w)\exp(-2i \arg w_z)} \quad .$$

Equation [8] has the form

$$\mu_{g \circ f} = (a+b)/(1+ab)$$

with a and b lying compactly in Δ except at the singularities. An easy estimate then gives $\mu_{g \circ f}$ also lies compactly in Δ, with the compact subset depending only on $\|\mu_f\|_\infty$ and $\|\mu_g\|_\infty$. Thus $\|\mu_{g \circ f}\|_\infty < 1$ and the proof is complete. ///

<u>Teichmüller's Uniqueness Theorem</u>

<u>If</u> $f:S \to S' = (S,\omega,k_0)$ <u>is an admissible diffeomorphism homotopic to the identity, then</u>

$$K[f] \geq K_0 = (1+k_0)/(1-k_0) \quad .$$

<u>Equality holds only if</u> $f = \mathrm{id}$.

<u>Proof</u>: Let f_0 denote $\mathrm{id}:S \to S'$ and $h = f \circ f_0^{-1}$. By the previous Lemma $h:S' \to S'$ is admissible and homotopic to the

identity. We note several obvious facts valid for $p \in S \backslash (b_f \cup b_\omega)$.

[9] $$\lambda_{f,\omega,\omega'} = K_0 \lambda_{h,\omega',\omega'}$$

[10] $$j_{f_0,\omega,\omega'} = K_0 \quad .$$

More important and somewhat subtler is

[11] $$\lambda^2_{f,\omega,\omega'} \leq K[f] j_{f,\omega,\omega'} \quad .$$

To prove [11], in local coordinates, we have

[12] $$K_f(\zeta) j_{f,\omega,\omega'} = \frac{|f_\zeta| + |f_{\bar\zeta}|}{|f_\zeta| - |f_{\bar\zeta}|} \left(|f_\zeta|^2 - |f_{\bar\zeta}|^2\right)$$

$$= \left(|f_\zeta| + |f_{\bar\zeta}|\right)^2$$

$$\geq \frac{1}{2}\left((\xi'_\xi)^2 + (\xi'_\eta)^2 + (\eta'_\xi)^2 + (\eta'_\eta)^2 + 2\text{Re}\, f_\zeta f_{\bar\zeta}\right.$$

$$= (\xi'_\xi)^2 + (\eta'_\xi)^2$$

$$= \lambda_{f,\omega,\omega'} \quad .$$

Since $K[f] \geq K_f(\zeta)$ on $S \backslash (b_f \cup b_\omega)$, [11] is proved.

We now have, using [9], [10] and [11],

$$\iint_{S'} \lambda^2_{h,\omega',\omega'} dA_{\omega'} = K_0^{-2} \iint_{S'} \lambda_{f,\omega,\omega'} dA_{\omega'}$$

$$= K_0^{-1} \iint_S \lambda_{f,\omega,\omega'} dA_\omega$$

$$\geq \frac{K[f]}{K_0} \iint_S j_{f,\omega,\omega'} dA_\omega$$

$$= \frac{K[f]}{K_0} \iint_{S'} dA_{\omega'} \quad .$$

Combining this result with the previous corollary, we have $K[f] \geq K_0$. If we have equality then, in particular,

$$[13] \qquad \lambda^2_{f,\omega,\omega'} = K_0 j_{f,\omega,\omega'}$$

almost everywhere, hence everywhere in $S\backslash(b_f \cup b_\omega)$. We write $\zeta = \xi + i\eta$ and

$$f:S \to S'$$
$$:\zeta \mapsto \zeta' = \xi' + i\eta'$$

in the ω'-coordinate. Then, with $k_0 = (K_0-1)/(K_0+1)$

$$|(\zeta')_{\bar\zeta} - k_0(\zeta')_\zeta|^2 =$$
$$(K_0+1)^{-2}[K_0^2((\eta'_\eta)^2+(\xi'_\eta)^2) - K_0(\xi'_\xi\eta'_\eta - \xi'_\eta\eta'_\xi)].$$

But by [2] and [13],

$$K_0^2((\eta'_\eta)^2+(\xi'_\eta)^2) \leq K_0 j_{f,\omega,\omega'} = K_0(\xi'_\xi\eta'_\eta - \xi'_\eta\eta'_\xi)$$

and

$$\zeta'_{\bar\zeta} = k_0\zeta'_\zeta \quad .$$

By Proposition (3.4), $f \mid (S\backslash(b_f \cup b_\omega)$ is conformal in the S' structure. Since f is continuous on S, f is conformal on S'. Since it is homotopic to the identity, by §1, $f = \mathrm{id}$. ///

<u>Corollary</u>: If $k_i \in]0,1[$ and $\omega_i \in \mathcal{Q}_S\backslash\{0\}$, then $S_1 = (S,\omega_1,k_1)$ is conformally equivalent to $S_2 = (S,\omega_2,k_2)$ via a map homotopic

to the identity if and only if $k_1 = k_2$ and $\frac{\omega_1}{\omega_2} \in \mathbb{R}^+$.

Proof: Sufficiency is trivial. To show necessity, we consider $id_i : S \to S_i$ and a conformal map $f : S_1 \to S_2$ homotopic to the identity. It is easily seen that $f \circ id_1 : S \to S_2$ is admissible and the Theorem gives

$$\frac{1 + k_1}{1 - k_1} = K[f \circ id_1] \geq \frac{1 + k_2}{1 - k_2} \quad .$$

Interchanging the roles of S_1 and S_2 we obtain $k_1 = k_2$ and $f \circ id_1 = id_2$. Thus purely as a set map $f = id$.

For any admissible F, μ_F has been defined in Lemma 2. If ζ_i is the ω_i-coordinate on S, locally $\zeta_2 = g(\zeta_1)$ where g is holomorphic. By equation [8],

$$k_1 = \mu_{id_1}(\zeta_1) = \mu_{f^{-1} \circ id_2 \circ g}(\zeta_1) = \mu_{id_2 \circ g}(\zeta_1)$$

$$= \mu_{id_2}(\zeta_2) \exp[-2i \arg(\zeta_2)_{\zeta_1}] \quad .$$

But

$$\mu_{id_2}(\zeta_2) = k_2 = k_1$$

and consequently

$$(\zeta_2)_{\zeta_1} \in \mathbb{R}\backslash\{0\}$$

or

$$\zeta_2 = c\zeta_1$$

for some $c \in \mathbb{R}\backslash\{0\}$. It follows immediately that

$$\omega_1 = d\zeta_1^2 = c^2 d\zeta_2^2 = c^2 \omega_2 \quad .$$

§4 Teichmüller's Existence Theorem

(4.1) Riemannian metrics on compact surfaces

A Riemannian metric on a manifold M is a positive definite symmetric bilinear form on the tangent bundle of M. It is preferable on surfaces S to think of Riemannian metrics as Gauss did, i.e., as a form

$$ds^2 = E\,dx^2 + 2F\,dxdy + G\,dy^2$$

with $EG - F^2 > 0$.

A tedious algebraic juggle gives the complex form

$$ds^2 = \lambda(z)\,|\,dz + \mu(z)d\bar{z}\,|^2$$

in terms of a local coordinate z. In this form $\lambda(z) > 0$ and $\|\mu(z)\|_\infty < 1$. Invariance under change of local coordinate requires that μ be a complex valued $(-1,1)$-form, i.e., $\mu d\bar{z}/dz$ is invariant. With respect to the local coordinate z, $\lambda(z)$ is a stretching factor constant with respect to changes of direction at z. λ is a $(1,1)$-form, i.e., $\lambda dz d\bar{z}$ is invariant. μ measures the directional change of infinitesimal distortion of distance.

If $ds_i^2 = \lambda_i\,|\,dz + \mu_i d\bar{z}\,|^2$ are Riemannian metrics on a surface S, then $id:(S,ds_1^2) \to (S,ds_2^2)$ is conformal whenever $\mu_1 = \mu_2$. The scale factors are thus irrelevant and the μ's may be used to parametrize conformal equivalence classes of Riemannian metrics. For more general diffeomorphisms we compute in the universal cover as follows.

Lemma 1: Let $f:U \to U$ be a diffeomorphism and $d\sigma^2 = y^{-2}(dz^2 + dy^2)$ denote the Poincaré metric in U. Suppose that $ds^2 = \lambda\,|\,dz + \mu d\bar{z}\,|^2$

is another Riemannian metric on U. Then the following are equivalent:

(i) f is conformal as a map $(U, ds^2) \to (U, d\sigma^2)$

(ii) f satisfies the <u>Beltrami equation</u>

$$f_{\bar{z}} = \mu f_z \quad .$$

<u>Proof</u>: f is conformal precisely when $|df|^2$ is independent of direction. If f satisfies the Beltrami equation then

$$|df|^2 = |f_z dz + f_{\bar{z}} d\bar{z}|^2 = |f_z|^2 |dz + \mu d\bar{z}|^2 = C(z) ds^2$$

where $C(z)$ depends only on z. Thus $|df|^2$ is independent of direction as a map from (S, ds^2). The computation also yields the reverse implication. ///

An easy computation shows that f covers a map $U/G \to U/G$ if and only if λ is a $(1,1)$-form for G and μ is a $(-1,1)$-form G, i.e., $\lambda(\gamma(z)) |\gamma'(z)|^2 = \lambda(z)$ and $\mu(\gamma(z)) \overline{\gamma'}(z)/\gamma'(z) = \mu(z)$ for all $\gamma \in G$. Every hyperbolic surface admits a smooth metric, namely the Poincaré metric. In terms of any local coordinate z on a Riemann surface S, we see that the Teichmüller deformation $S' = (S, \omega, k)$, with $\omega = \varphi(z) dz^2$, carries the singular metric

$$ds^2_{\omega,k} = |dz + \mu d\bar{z}|^2$$

where $\mu = k \ |\varphi|/\varphi$.

<u>Lemma 2</u>: If ζ is the ω-coordinate on $S \backslash b_\omega$ and $\zeta' = (\zeta + k\bar{\zeta})/(1-k)$ is the ω-coordinate on $S' = (S, \omega, k)$ then $ds^2_{\omega'} = C(z) ds^2_{\omega,k}$ where $ds_{\omega'}$ is the ω'-line element on S' and C is independent of direction.

<u>Proof</u>:

$$| d\zeta' |^2 = | (\zeta')_\zeta d\zeta + (\zeta')_{\bar\zeta} d\bar\zeta |^2$$

$$= | \varphi | (1-k)^{-2} | dz + k(| \varphi | /\varphi) d\bar{z} |^2$$

since $d\zeta = \varphi^{1/2} dz$. ///

The forms μ are called <u>Beltrami coefficients</u>. A Beltrami coefficient $\mu(z) = k(|\varphi|/\varphi)$ where $k \in]0,1[$ and $\omega = \varphi(z)dz^2 \in \mathcal{Q}_S \setminus \{0\}$, is called a <u>Teichmüller differential</u>.

(4.2) <u>Teichmüller's existence theorem</u>

We will next show that there is a natural topology on the set of Teichmüller deformations and a natural map Φ of these deformations which is a homeomorphism onto F_g. In particular every conformal structure on S may be obtained using a Teichmüller deformation of S.

Let $U = \{\operatorname{Im} z > 0\}$, S be a compact Riemann surface of genus $g > 1$ and $\tilde{S}$ be the holomorphic universal cover of S. Generically (i.e., possibly with subscripts) let $S' = (S, \omega, k)$ with $\omega \in \mathcal{Q}_S \setminus \{0\}$ and $k \in]0,1[$, and $\tilde{S}'$ be $\tilde{S}$ with the ω, k structure lifted. Using the uniformization theorem we obtain a diagram

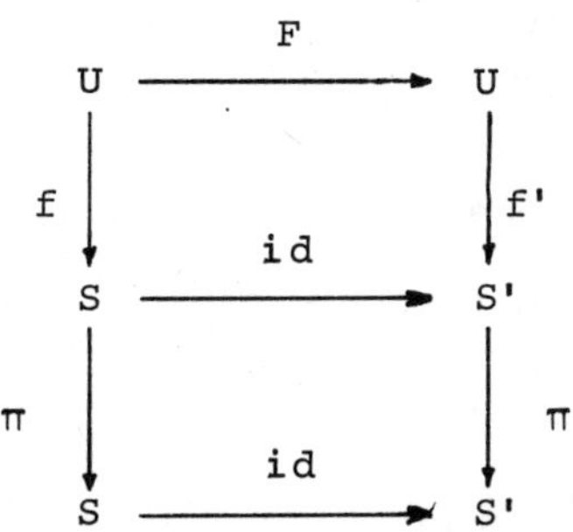

where f and f' are conformal and F is defined by the rest of the diagram. The group of deck transformations G (respectively G') for the projection $\pi \circ f$ (respectively $\pi' \circ f'$) is defined up to an automorphism $\eta \in \mathrm{Aut}\, U$. If $\pi_1 S = \langle a_1, b_1, a_2, b_2, \ldots, a_g, b_g : \prod_1^g [a_i, b_i] \rangle$ let $\gamma_1, \ldots, \gamma_{2g} \in G$ (respectively, $\gamma_1', \ldots, \gamma_2' \in G'$) be generators of G under a fixed isomorphism of G with $\pi_1 S$ as in §1 (respectively, under the isomorphism with G induced by id). We normalize G and G' so that 0 is the repelling and ∞ the attracting fixed point of γ_{2g} and 1 is a fixed point of γ_{2g-1}. These normalizations result in normalizations of F and f' and define a map

$$\begin{aligned} \Phi_1 : (\mathcal{Q}_S \setminus \{0\}) \times]0,1[&\to F_g \\ (\omega, k) &\mapsto G' \end{aligned}$$

where G' is the normalized group of deck transformations for $S' = (S, \omega, k)$. As we have seen, this map is not injective.

We norm $\mathcal{Q}_S$ by $\|\omega\| = \iint_S |\omega|$ and let $\mathcal{Q}_S^1$ be the open unit ball in $\mathcal{Q}_S$. Let

$$p : \mathcal{Q}_S^1 \to \{(\mathcal{Q}_S \setminus \{0\}) \times]0,1[\} \cup \{0\}$$

$$\omega \to \begin{cases} 0 & \text{if } \omega = 0 \\ (\omega/\|\omega\| \,,\, \|\omega\|) & \text{if } \omega \neq 0 \end{cases} .$$

p is a homeomorphism of the Teichmüller space $T(S)$ and $\mathcal{Q}_S^1$. Further let $\Phi : \mathcal{Q}_S^1 \to F_g$ be $\Phi_1 \circ p$.

<u>Lemma</u>: Φ is injective.

<u>Proof</u>: If $\Phi(\omega_1) = \Phi(\omega_2)$ then $(S, \omega_1/\|\omega_1\| \,,\, \|\omega_1\|)$ and $(S, \omega_2/\|\omega_2\| \,,\, \|\omega_2\|)$ are conformally equivalent via a map homotopic

to the identity. By the corollary to Teichmüller's uniqueness theorem, $\|\omega_1\| = \|\omega_2\|$ and $\omega_1/\omega_2 \in \mathbb{R}^+$, i.e., $\omega_1 = \omega_2$. ///

We have already mentioned that $\mathcal{Q}_S$ is a $3g - 3$ dimensional complex linear space. Using invariance of domain, we show that Φ is a homeomorphism by simply showing that it is a continuous surjection. We shall need the following theorem of Ahlfors and Bers.

<u>Theorem</u>: <u>Let</u> $\mu \in L^\infty(\mathbb{C})$ <u>satisfy</u> $\|\mu\|_\infty < 1$. <u>Then there exists a unique homeomorphism</u> $w^\mu : \mathbb{C} \to \mathbb{C}$ <u>so that</u>

(i) $(w^\mu)_{\bar{z}} = \mu(z)(w^\mu)_z$,

(ii) w^μ <u>fixes</u> $0,1$ <u>and</u> ∞ ,

(iii) w^μ <u>is smooth wherever</u> μ <u>is</u> ,

(iv) <u>for</u> $k = 0,1,\ldots,\omega$, w^μ <u>depends</u> $\mathcal{C}^k$ <u>or complex analytically on any parameters which</u> μ <u>does</u>.

<u>Proof</u>: See Ahlfors, [2], Chapter V.

Some remarks are in order about the previous theorem.

1) The derivatives taken in (i) are generally locally integrable distributional derivatives. This is unnecessary for our purposes.

2) Conclusions (i) and (ii), for gradually weaker conditions on μ, were obtained by Gauss, Korn and Lichtenstein, and for $\mu \in L^\infty$ by Morrey (1938). The main contribution of Ahlfors and Bers is the dependence on parameters. In particular, that the Teichmüller space is a complex manifold may be derived quite easily from (iv). Teichmüller directly - and erroneously - stated that he had only achieved a real analytic parametrization of F_g.

3) If μ satisfies $\overline{\mu}(\overline{z}) = \mu(z)$ then the uniqueness immediately implies that w^{μ} restricts to a homeomorphism of $\mathbb{R}$. It then follows that w^{μ} restricts to a self-map of U and $\overline{w^{\mu}}(\overline{z}) = w^{\mu}(z)$ for $z \in U$. In the following corollaries, if μ is defined on U we shall always assume that μ is extended to $\mathbb{C}$ by setting $\mu(\overline{z}) = \overline{\mu(z)}$.

Corollary 1: Φ is continuous.

Proof: Let

$$\Psi_1 : \mathcal{Q}_S^1 \to L^{\infty}(U)$$

$$: \omega \mapsto \begin{cases} 0 & \text{if } \omega = 0 \\ \mu_{\tilde{\omega}} = \|\tilde{\omega}\| \ \overline{(\tilde{\omega})}/(\tilde{\omega}) & \text{if } \omega \neq 0 \end{cases}$$

where $\tilde{\omega}$ covers ω. Since S is compact, convergence in the norm topology on $\mathcal{Q}_S$ implies convergence in the uniform topology. Thus, if $\omega_n \to \omega$ in $\mathcal{Q}_S^1$, $\tilde{\omega}_n$ converges to $\tilde{\omega}$ normally. But if $S = U/G$ and $\gamma \in G$,

$$\mu_{\tilde{\omega}}(\gamma(z))\overline{\gamma'(z)}/\gamma'(z) = \mu_{\tilde{\omega}}(z) \quad .$$

Thus the convergence of $\mu_{\tilde{\omega}_n}$ to $\mu_{\tilde{\omega}}$ is uniform everywhere (except at the zeroes of $\tilde{\omega}$ and $\tilde{\omega}_n$) and Ψ_1 is continuous.

Let $\Psi_2(\mu) = w^{\mu} \mid U$. If we give the autohomeomorphisms of U the uniform topology, then by the Ahlfors-Bers Theorem Ψ_2 is continuous.

Away from $\pi^{-1}(b_{\omega})$, $F \circ f^{-1}$ satisfies the Beltrami equation

$$(F \circ f^{-1})_{\overline{\zeta}} = \|\omega\| \ (F \circ f^{-1})_{\zeta}$$

by Proposition 3.4 where $F = \Phi(\omega)$. But ζ is a holomorphic function of $z \in \tilde{S}$. A short local computation yields

$$(F \circ f^{-1})_{\bar{z}} = \mu_\omega(z)(F \circ f^{-1})_z \quad .$$

By the uniqueness part of the theorem, $F = w^{\mu_\omega} | U$.

The matrix entries of a finite set of generators of $G' = F G F^{-1}$ are therefore continuous functions of $\omega \in \mathcal{Q}^1_S$ and Φ is continuous. ///

<u>Corollary 2</u>: Φ is a surjection onto a component of F_g.

<u>Proof</u>: Suppose $G' \in \partial\Phi(\mathcal{Q}^1_S)$ and $S = U/G$ as a marked surface. Let F be a differentiable map $F:U \to U$ covering $f:S \to U/G'$. Then $\mu(z) = F_{\bar{z}}/F_z \in L^\infty(U)$ with $\|\mu\|_\infty < 1$. Since the image of $\mathcal{C}^0(U)$ is dense in $L^\infty(U)$ and $\Phi(\mathcal{Q}^1_S)$ is open, there exists differentiable mappings F_n and f_n so that $\mu_{F_n} \to \mu$ and

$$\begin{array}{ccccc} U & \xrightarrow{F_n} & U & & \\ \downarrow & & \downarrow & & \\ S & \xrightarrow{f_n} & U/G_n & \xrightarrow{g_n} & S'_n = \left(S, \frac{\omega_n}{\|\omega_n\|}, \|\omega_n\|\right) \end{array}$$

We assume that f_n and g_n preserve markings and g_n is conformal. If μ_n is the Teichmüller differential associated to ω_n

$$K[f_n] = \frac{1 + \|\mu_n\|_\infty}{1 - \|\mu_n\|_\infty} < K < \infty$$

if $\mu_n \to \mu$. By Teichmüller's uniqueness theorem,

$$K[f_n] \geq \frac{1 + \|\omega_n\|}{1 - \|\omega_n\|} \to \infty$$

since $\omega_n \to \partial\mathcal{Q}^1_S$. This contradiction proves that $\partial\Phi(\mathcal{Q}^1_S) = \emptyset$ or

Φ is a surjection of $\mathcal{Q}_S^1$ onto a component of F_g. ///

Corollary 3: F_g is connected.

Proof: Let S_0 be a fixed marked surface, $S_0 = U/G_0$. Each $x \in F_g$ determines a monomorphism $\chi:\pi_1 S_0 \to \Gamma$. Set $G_x = \chi(\pi_1(S_0))$. $S_x = U/G_x$ is also a marked surface. A smoothed PL map gives a diffeomorphism $f:S_0 \to S_x$ which preserves markings. f lifts to $\tilde{f}:U \to U$ and $\mu_{\tilde{f}} = \tilde{f}_{\bar{z}}/\tilde{f}_z \in L^\infty(U)$ satisfies $\|\mu_{\tilde{f}}\|_\infty < 1$. Let w be the normalized solution to the Beltrami equation $w_{\bar{z}} = \mu_{\tilde{f}} w_z$. By the uniqueness of the normalized solution to the Beltrami equation $w = \tilde{f}$.

Let $\mu_t = t\mu_{\tilde{f}}$ and w_t be the normalized solution to $(w_t)_{\bar{z}} = \mu_t (w_t)_z$. By computing z derivatives we see that $G_t = w_t G_0 (w_t)^{-1} \subset \Gamma$ and the map

$$
\begin{aligned}
[0,1] &\to F_g \\
t &\mapsto x_t
\end{aligned}
$$

where x_t is defined by $G_{x_t} = G_t$, is continuous. F_g is therefore path connected. ///

We have proved

Theorem: (Teichmüller) $\Phi:\mathcal{Q}_S^1 \to F_g$ is a homeomorphism.

Corollary 1: (Teichmüller's Existence Theorem) If S is a compact surface of genus $g > 1$, then every complex structure on S, viewed as a marked surface, may be obtained using a Teichmüller deformation.

The corollary should be interpreted as meaning that every deformation of structure is homotopic to a Teichmüller deformation (or is trivial). The proof of the Teichmüller theorems is now complete.

Another immediate consequence of the Theorem is

Corollary 2: Let $f:S_1 \to S_2$ be a homeomorphism. Then in the homotopy class of f there exists a map $F = h \circ f_1$ where f_1 is a Teichmüller deformation of S_1 and h is a conformal map.

The mapping F of Corollary 2 is called the Teichmüller map in the class of f.

§5 Teichmüller Space

Our construction thus far has relied heavily on the choice of a base surface S for $\mathcal{Q}^1_S$. Teichmüller also showed that there is a metric space structure on $\mathcal{Q}^1_S$ parametrizing the conformal structures on the marked surface S. The metric does not (by definition) depend on the choice of base surface. Let g be the genus of S.

We identify points in $\mathcal{Q}^1_S$ with the surfaces they represent. For $S', S'' \in \mathcal{Q}^1_S$ let

$$d_T(S', S'') = \log K$$

where $S'' = (S', \omega/\|\omega\|, \|\omega\|)$, $\omega \in \mathcal{Q}^1_{S'}$, and $K = (1+\|\omega\|)/(1-\|\omega\|)$. The Teichmüller space T_g is the metric space $(\mathcal{Q}^1_S, d_T)$.

Theorem: T_g is a complete metric space homeomorphic to F_g and $\mathcal{Q}^1_S$.

Proof: That d_T is a metric follows from §3.5 and Teichmüller's uniqueness theorem. If $S_n \in T_g$ and (S_n) is Cauchy in d_T then (S_n) is Cauchy in $\mathcal{Q}^1_S$ hence T_g is complete. If $S_n = U/G_n$ and $d_T(S_n, S_0) \to 0$, it is easy to see that $G_n \to G_0$ and conversely.

///

Kravetz has proved many interesting results on the geometry of Teichmüller spaces - the incorrect ones known to the author are discussed in the Columbia University dissertation of M. Linch, and H. Masur's Geodesics in Teichmüller Space.

§6 Historical Notes

The Fricke coordinates which we have established are not precisely those due to Fricke. He used a distinguished set of curves whose lengths were the parameters. This can be related to the traces of the matrices covering the curves (c.f. the next Chapter). It is not clear that Fricke had proofs of his results in the modern mathematical sense. His treatment is close to unintelligible, indeed one is never sure of what he claims to prove. Using the techniques of quasiconformal mapping, L. Keen [7] has developed a theory of Fricke coordinates or moduli.

Teichmüller's papers are also generally considered unreadable. The first acceptable proof of the Teichmüller theorems was given by Ahlfors [1]. The proof given here follows Bers [4], although the proof of the uniqueness theorem is essentially Teichmüller's.

An analytic geometric interpretation of the Teichmüller theorems may be found in Hamilton [5] and Hubbard [6].

REFERENCES

[1] L. Ahlfors, On quasiconformal mappings, J. d'Analyse Math. 4 (1954), 1-58.

[2] __________, Lectures on Quasiconformal Mappings, Van Nostrand, 1966.

[3] __________, Conformal Invariants, McGraw-Hill, 1973.

[4] L. Bers, Quasiconformal mappings and Teichmüller's Theorem, Analytic Functions, Princeton Univ. Press, (1960), 89-119.

[5] R. Hamilton, Extremal quasiconformal mappings with prescribed boundary values, Trans. Amer. Math. Soc. 138 (1969), 399-406.

[6] J. Hubbard, Sur les sections analytiques de la courbe universelle de Teichmüller, Memoirs of the Amer. Math. Soc. 166 (1976).

[7] L. Keen, On Fricke moduli, Ann. of Math. Studies 66 (1971), 205-224.

CHAPTER II

TOPICS IN TEICHMÜLLER THEORY

The previous Chapter presented the Teichmüller theorems for compact surfaces. The only result there not known in the 1940's is the Ahlfors-Bers theorem. In this Chapter, we have two goals in mind. First, we construct a compactification $\hat{R}(S_0)$ of the Riemann space $R(S_0)$ of a conformally finite Riemann surface S_0. The points of $\hat{R}(S_0) \setminus R(S_0)$ represent Riemann surfaces with nodes; i.e., on the topological surface S_0, a finite set of homotopically independent disjoint simple closed curves have been pinched to points. Our second objective is to lay the groundwork for the third chapter, in which the diffeomorphisms of surfaces are classified and continuous group actions on Teichmüller space are studied.

Our first section is essentially due to Teichmüller. We give the Teichmüller theorems for topologically finite Riemann surfaces using techniques due to Ahlfors. In §2, the Teichmüller modular group is defined and shown to act discontinuously on the Teichmüller space. §3 is devoted to augmented moduli spaces and the Fenchel-Nielsen coordinates on them. It concludes with a proof of the compactness of $\hat{R}(S_0)$.

§1 Punctured surfaces with boundary and their Teichmüller spaces

A surface S is *topologically* (or *differentiably*) *finite* if $\pi_1 S$ is finitely generated or, equivalently, there is a compact surface S' and a continuous (or differentiable) injection $i: S \to S'$ so that $S' \setminus i(S)$ is finite. The *genus* of S is the genus of S'. The genus of S and the number of points in $S' \setminus i(S)$ classifies the homeomorphism or diffeomorphism classes of orientable finite surfaces. When S has conformal structure, we may distinguish between two kinds of ideal boundary components, i.e., points of $S' \setminus i(S)$. Let $z \in S' \setminus i(S)$ and N' be a small disc about z so that $N = N' \setminus \{z\} \subset i(S)$. Using the uniformization theorem, we see that $i^{-1}(N)$ is either a punctured disc or an annulus. In the first case, z is called a _puncture_. S has type (g,n,m) if S has genus g, $S' \setminus i(S)$ has $n + m$ points and n of these points are punctures. S is _conformally finite_ if S is topologically finite and $m = 0$. All surfaces we consider here are orientable.

(1.1) _Bordered surfaces_

Let S be a surface of type (g,n,m) with $m > 0$. We wish to show that S may be viewed as the interior of a bordered surface with m border curves. Our proof, which is based on the Schwarz Lemma, yields several other results which will be useful in our later discussion.

Let S_1 and S_2 be hyperbolic Riemann surfaces with $S_1 \subset S_2$ and universal cover Δ. S_i inherits a complete hyperbolic metric, of course with constant negative curvature, from the covering projection $\pi_i : \Delta \to S_i$. In the local coordinate z near $\zeta_0 \in S_1$, we may write this metric as

[1] $$ds_i^2 = \lambda_i(z)\,|\,dz\,|^2.$$

ds_i^2 transforms as a $(-1,1)$-form, i.e., if z' is another local coordinate at ζ_0 and $ds_i^2(z') = \lambda_i'(z')\,|\,dz'\,|^2$, then

[2] $$\lambda_i'(z') = \lambda_i(z)\ |\ dz/dz'\ |^2.$$

It is now easy to show

Theorem (the Generalized Schwarz Lemma)

If $S_1 \subset S_2$ *and* S_2 *is hyperbolic, then*

$$\lambda_1(\zeta) \geq \lambda_2(\zeta)$$

with equality if and only if $S_1 = S_2$.

Proof: By the classification of non-hyperbolic Riemann surfaces, we see that $\tilde{S}_1 = \Delta$. Let $\zeta_0 \in S_1$ and assume $\pi_i(0) = \zeta_0$, if necessary by translating by an element of $\text{Aut}\ \Delta$. We may find a holomorphic injection F which commutes the diagram.

$$\begin{array}{ccc} \Delta & \xrightarrow{F} & \Delta \\ \pi_i \downarrow & & \downarrow \pi_2 \\ S_1 & \xrightarrow{i} & S_2 \end{array}$$

with $F(0) = 0$. The classical Schwarz Lemma shows that $|\ F'(0)\ | \leq 1$. If σ denotes the hyperbolic metric in Δ, set $\sigma_i = \pi_{i*}(\sigma)$. Near $z = 0$, $\pi_i(z)$ is a local coordinate near ζ_0, hence σ_i may be written

$$ds_i^2 = \sigma_i = \lambda_i(\pi_i(z)\ |\ d\pi_i(z)\ |^2.$$

If λ_1' is the coefficient of ds_1^2 in the local coordinate π_2, from [2] we obtain

$$\lambda_1'(\pi_2) = \lambda_1(\pi_1) \left|\frac{d\pi_1}{d\pi_2}\right|^2 = \lambda_1(\pi_1)\ |\ F'\ |^{-2} \geq \lambda_1(\pi_1).$$

Evaluate at 0 to obtain

$$\lambda_1'(\pi_2) \geq \lambda_1(\pi_1) = \lambda_2(\pi_2).$$

For equality, F, hence also i, must be surjective. This proves the theorem for the local coordinate $\pi_2(z)$ at ζ_0. ζ_0 was chosen arbitrarily and change of local coordinate multiplies both sides of the inequality by the same factor. The proof is complete. ///

Corollary 1: Let ζ_0 be a puncture on S. The deck transformation γ of Δ covering a small simple loop C about ζ_0 is parabolic.

Proof: Let N be a small neighborhood of ζ_0 which is conformally a punctured disc and $\tilde{N}$ be the universal cover of N. We identify $\tilde{N}$ with the upper half-plane U.

$$\pi: U \to N$$
$$z \mapsto e^{-i\log z}$$

is the universal cover map. If ds_1^2 is the hyperbolic metric in N then the length of the curve C on N is $\int_C ds_1$. A simple loop around the puncture is covered by an open curve from some point $x + iy$ to $(x+2\pi) + iy$. The length of the curve in the hyperbolic metric thus goes to zero as the curve is retracted into the puncture. The length of C in the hyperbolic metric ds_2^2 on S is given by $\int_C ds_2 \leq \int_C ds_1$. Thus the infimum over the homotopy class of C of the ds_1^2 lengths of curves C' is zero.

If $\pi_1: U \to S$ is the universal cover map and γ_1 is a deck transformation covering C, we show that γ_1 cannot be hyperbolic.

If γ_1 were a hyperbolic transformation, we may assume that its fixed points lie at 0 and ∞. C lifts to an arc going from a point z_0 to $\gamma_1(z_0)$. The hyperbolic metric in U is $ds^2 = y^{-2}(dx^2 + dy^2)$, hence the infimum is attained along the vertical line through 0 and is non-zero. ///

The converse is also true, namely

Corollary 2: Let ζ_0 be an ideal boundary component of S which is not a puncture. The deck transformation γ_1 of Δ covering a small loop C about ζ_0 is hyperbolic.

Proof: Since C separates S any component $\tilde{C}$ of the lift of C to U must separate U. If γ_1 were parabolic, it would have a single fixed point, say ∞. In one component of $U \setminus \tilde{C}$, the covering map would again be the exponential and ζ_0 would be a puncture. ///

Implicit in the above proofs is the assertion that a homotopy class determined by a parabolic deck transformation has no geodesic representative. A class determined by a hyperbolic deck transformation γ contains a geodesic. The geodesic is the projection of the hyperbolic line L through the fixed points of γ.

Let C be a simple closed curve on S surrounding an ideal boundary component ζ_0 which is not a puncture and assume that C determines the deck transformation γ for the covering $\pi: U \to S$. C lifts to a γ-invariant curve $\tilde{C}$ connecting the fixed points of γ. One of the components B of $U \setminus \tilde{C}$ projects to a neighborhood N of ζ_0 which is conformally an annulus, in fact $N = B/\langle\gamma\rangle$. The boundary of B in $\hat{\mathbb{C}}$ contains a closed subarc $\bar{A}$ of $\partial\Delta$, whose endpoints are fixed points of γ. The cover group G acts discontinuously on the orbit GA of $A = \overline{A} \setminus \{\text{endpoints}\}$. $G(B \cup A)/G$ is a surface with boundary. For each ideal boundary component of S which is not a puncture we may find such such a border component. The resulting surface with boundary is called a bordered Riemann surface. More generally a bordered Riemann surface is a surface with boundary whose interior S^o has the structure of a Riemann surface. We call $S \setminus S^o$ the border of S.

A finite bordered Riemann surface is often compactified by filling in the punctures. The additional points are called distinguished points. In the current literature, punctures and ditinguished points are terms used interchangeably.

(1.2) The double of a surface and Nielsen extensions

Let S be a bordered Riemann surface and $\overline{S}$ be the mirror image of S. The (Schottky) double S^d of S is $S \cup \overline{S}$ with corresponding points on the border identified. More precisely, to each $\zeta \in S$, assign a point $\overline{\zeta} = i(\zeta)$. If $\zeta \in S^o$, let $S^o \supset N \xrightarrow{z_\alpha} \mathbb{C}$ be a chart at ζ. As chart at $\overline{\zeta}$ take $\overline{N} = \{\overline{\zeta} \mid i^{-1}(\overline{\zeta}) \in N\}$ with local coordinate $\overline{z}_\alpha = J \circ z_\alpha \circ i^{-1}$ where $J\colon \mathbb{C} \to \mathbb{C}$ is complex conjugation. For $\zeta \in \partial S$, we use the obvious topology. This defines the mirror image bordered surface $\overline{S}$ of S. Then $S^d = S \cup_i \overline{S}$.

The conformal structure of S^d is well defined on $S^o \cup \overline{S}^o$. Let N be a neighborhood of $\zeta \in S \setminus (S^o \cup \overline{S}^o)$ and f be continuous in N and holomorphic in $N_1 = N \cap (S^o \cup \overline{S}^o)$. As in (1.1), N may be injected into $\mathbb{C}$ via a map g with $g \mid N_1$ holomorphic and $g(N\backslash N_1) \subset \mathbb{R}$. Since straight lines are removable singularities for holomorphic mappings with continuous extensions, (c.f. Cartan [3], p. 71,74) $f \circ g^{-1}$ has a holomorphic extension to $g(N)$. It follows that f extends holomorphically to N and S^d has a unique conformal structure at ζ. Thus S^d is an unbordered Riemann surface.

A painless method for obtaining the double of a Riemann surface is as follows. Let S be the bordered surface, $\pi: U \to S^o$ be the universal cover map and G be the group of deck transformations for the universal covering π. We may assume that ∞ is a fixed point of some element of G. G acts discontinuously on an open subset Ω of $\mathbb{C}$. $\Omega \supset U \cup L$, where L is the lower half-plane. Also, G acts discontinuously on open subintervals of $\mathbb{R}$. $S^d = \Omega/G$. The map i is covered by J, i.e., there is a canonical anti-conformal involution of S^d which interchanges S and $\bar{S}$ and fixes all points of ∂S.

Let S be a bordered Riemann surface, that is, S is of type (g,n,m) with $m > 0$. If $(g,n,m) \neq (0,0,2)$ or $(0,1,1)$ then S^d is hyperbolic. i is covered by an isometry of the the universal cover of S^d. Thus each border component of S is the geodesic in its free homotopy class on S^d. To see this, simply compute which curves in Δ are left fixed by an anti-conformal isometry of period two and project them.

<u>Definition</u>: Let S be a bordered Riemann surface and S^d be its double. Assume S^d is hyperbolic. The restriction of the hyperbolic metric on S^d to S is called the <u>intrinsic metric</u> on S.

Our preceding discussion has demonstrated

Lemma 1: In the intrinsic metric, border curves are geodesic.

If we represent $\tilde{S}^d$ as U with cover group G^d, $\pi_1 S$ injects into G^d and its image G' stabilizes a simply connected component U' of $U \setminus \pi^{-1}(\partial S)$. $S = U'/G'$ is a deformation retract of $S^N = U/G'$. S^N is called the Nielsen extension of S (see Bers [2]). The Nielsen extension attaches infinite horns to S along the border curves and thereby completes the intrinsic metric. (see Figure 1).

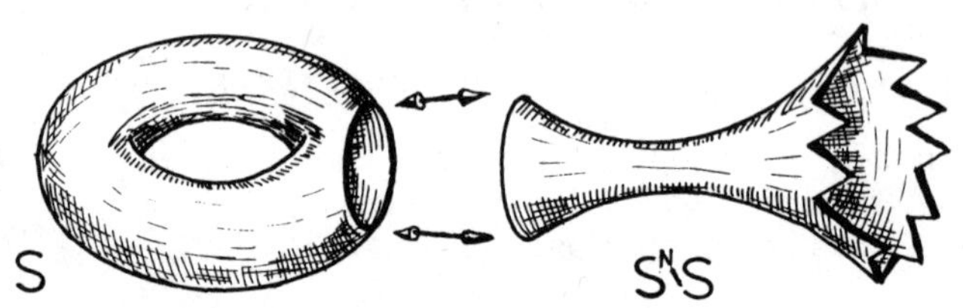

Figure 1

An elementary argument shows that any element $\gamma \in G^d$ covering a border curve of S is hyperbolic. Thus, by Corollary (1.1.1), $S_1 = S^N$ has border curves corresponding to each of its ideal boundary components. The Nielsen extension process then may be applied to S_1 to obtain S_2, etc. The border curves on S_j are geodesics in the (complete) hyperbolic metric on S_{j+1}.

Definition: The infinite Nielsen extension of S is

$$S_\infty = S_1 \cup S_2 \cup \dots \ .$$

Yet another application of the generalized Schwarz Lemma is

Lemma 2: Let C be a border curve of S and C_1 the border curve of the Nielsen extension S_1 of S freely homotopic to C. If ℓ is the length of C in the hyperbolic metric ds_1 on S_1 and ℓ_1 is the length of C_1 in the hyperbolic metric ds_2 on S_2, then

$$\ell_1 < \ell.$$

Proof: We have seen that $S_1 \subset S_2$ but $S_1 \neq S_2$. It then follows from Theorem (1.1) that $ds_1 > ds_2$. Further C_1 is the geodesic (with respect to ds_2) in the free homotopy class of C. Thus

$$\ell_1 = \int_{C_1} ds_2 < \int_C ds_2 < \int_C ds_1 = \ell. \qquad ///$$

(1.3) Annuli and punctured discs - the method of extremal length

Let S be a bordered Riemann surface homeomorphic to a punctured disc $\{0 < |z| \leq 1\}$. S then has type $(0,0,2)$ or $(0,1,1)$. S is always hyperbolic. In the first case, the cover group $G = \{z \mapsto \lambda^n z : n \in \mathbb{Z}\}$ for some real $\lambda > 1$. The second case corresponds to $G = \{z \mapsto z+n : n \in \mathbb{Z}\}$. We focus our attention on type $(0,0,2)$. The number λ may be used to parametrize the holomorphic equivalence classes of surfaces of type $(0,0,2)$. The covering projection

$$f(z) = \exp\{-2i(\log z)/\log \lambda\}$$

maps U into the annulus $\{\exp(2\pi^2/\log \lambda) < |z| < 1\}$. Alternatively the equivalence classes may be parametrized by $(2\pi)^{-1}\log (r_1/r_2)$ when S is conformally equivalent to the annulus $A = \{r_1 < |z| < r_2\}$. The module of A is the number $\mathfrak{m}(A) = (2\pi)^{-1}\log r_1/r_2$. An alternative description of the module of an annulus is given by the method of extremal length due to Ahlfors and Beurling.

Theorem 1: Let A be the annulus $\{r_1 < |z| < r_2\}$ and $R = \{\rho : A \to \mathbb{R}^+ \cup \{0\} \mid \rho$ is measurable and $0 < \iint_A \rho^2 dxdy < \infty\}$. Let Γ be the set of simple closed curves separating the border components of A and

$$\ell_\alpha(\rho) = \begin{cases} \int_\alpha \rho \, |dz| & \text{if } \rho \text{ is measurable on } \alpha \\ \infty & \text{otherwise} \end{cases}$$

If $L(\rho) = \inf\{\ell_\alpha(\rho) \mid \alpha \in \Gamma\}$, then

[3] $$\mathfrak{m}(A)^{-1} = \sup_{\rho \in R} \frac{L(\rho)^2}{\iint_A \rho^2 dxdy} .$$

Proof: In polar coordinates we have

$$L(\rho) \leq \int_0^{2\pi} \rho(re^{i\theta}) r d\theta$$

$$\frac{L(\rho)}{r} \leq \int_0^{2\pi} \rho d\theta .$$

Integrating with respect to r, we obtain

$$L(\rho) \log r_1/r_2 \leq \iint_A \rho dr d\theta.$$

The Schwarz Inequality then gives

$$L(\rho)^2 \log^2 r_1/r_2 \leq 2\pi \log (r_1/r_2) \iint_A \rho^2 r dr d\theta$$

or

$$\frac{L(\rho)^2}{\iint_A \rho^2 dxdy} \leq \mathfrak{m}(A)^{-1}.$$

That the inequality is sharp follows from considering the function $\rho = (2\pi r)^{-1}$. ///

If $F: S \to A$ is conformal and injective, we may pull back the curve family Γ and the densities $\rho \in R$. Since $\mathfrak{m}(A)$ is a conformal invariant so is the right-hand side of Equation [3]. This observation gives the Inequality of Grötzsch which is our

<u>Theorem 2</u>: (superadditivity of the module)
<u>Let</u> A_0 <u>be an annulus and</u> C <u>a simple closed curve separating the boundary components of</u> A_0. <u>If</u> A_1 <u>and</u> A_2 <u>are the two components of</u> $A_0 \setminus C$, <u>then</u>

$$\mathfrak{m}(A_0) \geq \mathfrak{m}(A_1) + \mathfrak{m}(A_2).$$

<u>Proof</u>: Let Γ_i and R_i be the curve family and set of densities respectively for A_i, $i = 1,2$. Clearly $\Gamma_0 \supset \Gamma_1 \cup \Gamma_2$. For $\rho \in R_0$, set

$$\rho_i(z) = \begin{cases} \rho \mid A_i & \text{for } z \in A_i \\ 0 & \text{for } z \in A_0 \setminus A_i \end{cases}$$

Let $L_i(\rho) = \inf\{\ell_\alpha(\rho_i) \mid \alpha \in \Gamma_i\}$. Then $L_i(\rho) \geq L_0(\rho)$ for $i = 1,2$ and

$$\iint_{A_0} \rho^2 dxdy = \iint_{A_1} \rho_1{}^2 dxdy + \iint_{A_2} \rho_2{}^2 dxdy.$$

Thus

$$\frac{\iint_{A_0} \rho^2 dxdy}{L_0(\rho)^2} \leq \frac{\iint_{A_1} \rho_1{}^2 dxdy}{L_1(\rho_1)^2} + \frac{\iint_{A_2} \rho_2{}^2 dxdy}{L_2(\rho_2)^2} .$$

The infimum of the left hand side is precisely $\mathfrak{m}(A_0)$. For $i = 1,2$, each $\rho \in R_i$ may be extended to a density $\rho_0 \in R_0$ by setting $\rho_0(z) = 0$ for $z \in A_0 \setminus A_i$. It follows that

$$\mathfrak{m}(A_i) = \inf_{\rho \in R_0} \frac{\iint_{A_i} \rho_i{}^2 dxdy}{L_i(\rho_i)^2}$$

and the conclusion follows immediately by taking infima over R_0 in the above inequality. ///

We may now apply these theorems to the Nielsen extension process. Let S be a bordered Riemann surface of type (g,n,m) with $m > 0$ and S^N be the Nielsen extension of S. The restriction of a complete hyperbolic metric on S^N to S is the intrinsic metric on S. Consequently each component of $S^N \setminus S$ is topologically an annulus A by Lemma 1.2.2. We compute $\mathfrak{m}(A)$.

Lemma: $\mathfrak{m}(A) = \pi^2/\log \lambda$ where $\log \lambda$ is the length of $\partial S \cap \partial A$ in the hyperbolic metric on S^N.

Proof: Let $\pi: U \to S^N$ be the universal cover map and $\gamma: z \mapsto \lambda z$ be a deck transformation covering $\partial S \cap \partial A$ oriented so that $\lambda > 1$ and $A = \pi(U \cap \{\text{Re } z > 0\})$. Then the computation preceding Theorem 1 shows that $\mathfrak{m}(A) = \pi^2/\log \lambda$. ///

<u>Theorem 3</u>: <u>If</u> S <u>is a bordered Riemann surface of type</u> (g,n,m) <u>with</u> $m > 0$ <u>then the infinite Nielsen extension</u> S_∞ <u>has no border curves, i.e.,</u> S_∞ <u>is a conformally finite surface</u>.

<u>Proof</u>: It suffices to show that each component A of $S_\infty \setminus S$ is conformally a punctured disc. To do so, we show that $\mathfrak{m}(A) = \infty$. Since $A = \bigcup [(S_i \setminus S_{i-1}) \cap A]$

$$\mathfrak{m}(A) \geq \sum ((S_i \setminus S_{i-1} \cap A)$$

$$= \sum \log \lambda_i$$

where $\log \lambda_i$ is the length of $\partial S_{i-1} \cap A$ in the intrinsic metric on S_i. By Lemma 1.2.2, $\log \lambda_i$ is an increasing sequence and $\mathfrak{m}(A) = \infty$. ///

We next examine the deformations of conformal structure of annuli induced by admissible mappings.

<u>Lemma 2</u>: Let A_1 and A_2 be annuli and $f: A_1 \to A_2$ be a diffeomorphism with isolated singularities. If $K = K[f] < \infty$, then

$$\mathfrak{m}(A_1) \leq K\, \mathfrak{m}(A_2).$$

<u>Proof</u>: Form the double A_i^d of A_i. A_i^d is a torus, hence has a period parallelogram P_i. f lifts to a map $F: P_1 \to P_2$ for suitably chosen P_2. From the solution to the Grötzsch problem, given in the previous Chapter, F minimizes $K[F_1]$ among all admissible maps $F_1: P_1 \to P_2$ precisely if F is affine.

Up to conformal equivalence, $A_i = \{1 < |z| < r_i\}$. Via the logarithm, map the set $(A_i \setminus \mathbb{R})$ into $\mathbb{C}$. An extremal admissible map $f_0: A_1 \to A_2$ is given by $\exp \circ g \circ \log$ where $g(x,y) = (px,y)$ and $p = (\log r_2)/(\log r_1)$. Since the logarithm is conformal on $A_i \setminus \mathbb{R}^+$,

$$K[f_0] = K[g] = \max\ (p, p^{-1}).$$

The conclusion of the lemma then follows from the fact that $K \geq K[f_0]$. ///

<u>Theorem 4</u>: (Wolpert) <u>Assume</u> S_1 <u>and</u> S_2 <u>are hyperbolic,</u> $f: S_1 \to S_2$ <u>is a diffeomorphism with isolated singularities and</u> $K[f] < \infty$. <u>If</u> α_1 <u>is a closed hyperbolic geodesic on</u> S_1 , <u>then the hyperbolic geodesic</u> α_2 <u>in the class of</u> $f(\alpha_1)$ <u>satisfies</u>

$$\ell(\alpha_2) \leq K[f]\ \ell(\alpha_1).$$

<u>Proof</u>: Let G_i be the group of deck transformations for the universal cover maps $\pi_i: U \to S_i = U/G_i$. By conjugation of G_i we may assume that α_i is covered by the geodesic from 0 to ∞. For $\lambda_i > 1$, let $\gamma_i: z \mapsto \lambda_i z$ be determined by α_i. f is covered by an admissible map $F: U \to U$ and $\gamma_2 = F \circ \gamma_1 \circ F^{-1}$. Let H_i be the cyclic group generated by γ_i. F covers a map $\psi: U/H_1 \to U/H_2$. U/H_i is an annulus and by the previous lemma

[4] $$\mathfrak{m}(U/H_1) \leq K[\psi]\ \mathfrak{m}(U/H_2).$$

Since astigmatism is a purely local property, $K[\psi] = K[F] = K[f]$. A simple computation, given in §2.2, Lemma 2, shows that $\ell(\alpha_i) = \log \lambda_i$. By the remarks at the beginning of §1.3,

$$\mathfrak{m}(U/H_i) = (2\pi)^{-1} \frac{2\pi^2}{\log \lambda_i} = \frac{\pi}{\ell(\alpha_i)}.$$

We insert this last equality into [4] to obtain the desired inequality.

///

(1.4) Teichmüller deformations of surfaces of finite topological type

Let S_0 be a Riemann surface of type (g,n,m). In order that the double of S_0 carry a hyperbolic metric, we assume that $(g,n,m) \neq (0,0,1)$, $(0,1,0),(0,0,1),(0,1,1),(0,2,0)$ or $(0,0,2)$. A Teichmüller theory of deformations of conformal structure of S_0 is an easy consequence of the Teichmüller theory for compact non-bordered surfaces once we have established the proper framework.

Definition: An admissible quadratic differential ω on S_0 is the restriction to S_0 of a meromorphic quadratic differential ω^d on S_0^d satisfying:

(i) at each of the n punctures on S_0, ω^d has at worst a first order pole,

(ii) ∂S_0 is an ω^d-horizontal line.

We denote by $\mathcal{Q}_{S_0}$ the space of admissible quadratic differentials on S_0.

A more geometric difinition is that ω is real on ∂S_0 and has at each puncture a simple pole or extends holomorphically across that puncture. Near a puncture ζ_0 at which ω has a pole, the horizontal (or vertical) foliation has the form shown in Figure 2. The possible types of leaf structures near a border curve are shown in Figure 3. Along the curve A, ω^d has no zeroes; along B ω^d has one zero of order four. Notice that symmetry requires that the zeroes of ω^d on ∂S_0 have even order.

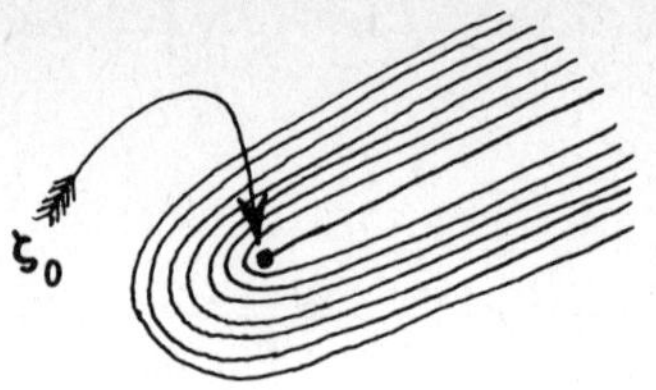

Figure 2

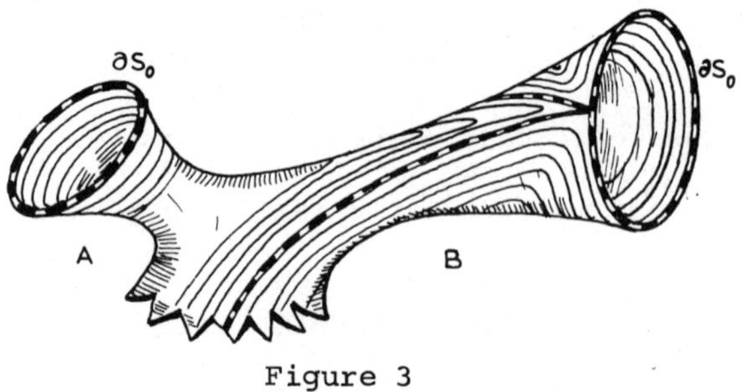

Figure 3

As in the case of compact surfaces, away from the zeroes of ω, there is a local coordinate

$$\zeta = \xi + i\eta = \sqrt{\omega}\ .$$

If $\zeta' = \xi' + i\eta' = K\xi + i\eta$ then, as before, we may define a new bordered Riemann surface structure $S_0' = (S_0, \omega, k)$ on S_0. Here $k = (K-1)/(K+1)$. This is the <u>Teichmüller deformation</u> of S_0 given by ω and k.

<u>Definition</u>: Let S_0' be a structure on S_0 of a Riemann surface of finite type. $id: S_0 \to S_0'$ is <u>admissible</u> if:

(i) punctures are mapped to punctures

(ii) $id \mid S_0^{\circ}$ is a diffeomorphism except at a discrete set of points accumulating only at border curves

(iii) $\text{ess sup}\,\{K_f(\zeta) \mid \zeta \in S_0^{\circ}\} < \infty$

where in the local coordinates ζ and $\zeta' = f(\zeta)$

$$K_f(\zeta) = \frac{\left|\frac{\partial f}{\partial \zeta}\right| + \left|\frac{\partial f}{\partial \bar{\zeta}}\right|}{\left|\frac{\partial f}{\partial \zeta}\right| - \left|\frac{\partial f}{\partial \bar{\zeta}}\right|} .$$

It is implicit in the definition that border curves are mapped to border curves. Thus admissible deformations preserve the type of ideal boundary components.

Lemma 1: Teichmüller deformations are admissible.

Proof: It is only necessary to show that punctures are mapped to punctures. Let D be a small neighborhood of a puncture ζ_0. Let D' denote the S_0'-conformal structure on D. As in the beginning of Section 4.2 of Chapter I, there is a map $F: U \to U$ covering $id: D \to D'$. The group of deck transformations for the covering $\pi: U \to D$ is cyclic parabolic by Corollary (1.1.1). To show that D' is a punctured disc, it suffices, by Corollary (1.1.2), to show that $G' = FGF^{-1}$ is cyclic parabolic.

For a Teichmüller deformation $S_0' = (S_0, \omega, k)$, K_f is essentially constant, hence $\mu_F = F_{\bar{z}}/F_z$ satisfies $\|\mu_F\| < 1$. By the Ahlfors-Bers Theorem, F is the unique (up to composition with a Möbius transformation) solution to the Beltrami equation $w_{\bar{z}} = \mu_F w_z$. F therefore extends to a homeomorphism of $\bar{U}$. Thus any element of $G' \setminus \{id\}$ has a single fixed point, i.e., is parabolic. ///

(1.5) Teichmüller's theorem for surfaces of finite topological type

The Teichmüller theorems for compact unbordered surfaces may be extended to surfaces of finite topological type. In fact, the proof is an application of the theorems for surfaces of type $(g,0,0)$ where $g > 1$.

Let S_1 be a surface of type $(g,2n,0) \neq (0,2,0),\ (0,4,0)$ and S_2 be another Riemann surface structure on S_1. Further assume that $id: S_1 \to S_2$ is admissible. Using the classical (that is, due to Riemann) method of branch cuts, we may construct a two-sheeted covering S_1' of S_1, which is branched of order 2 at each of the punctures. id then lifts to a map $f: S_1' \to S_2'$ which makes the following diagram commute:

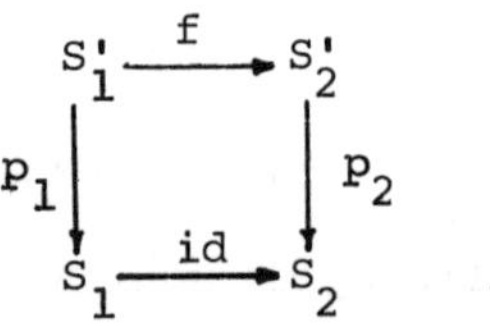

By the Teichmüller theorems for compact unbordered surfaces, among all admissible maps $f: S_1' \to S_2'$, the Teichmüller deformation is the unique map with minimal astigmatism K_1. Since admissible maps may be lifted with astigmatism preserved, $K[id] \geq K_1$. Equality holds if and only if f is the Teichmüller deformation f_0.

Let j_k denote the sheet interchange on S_k. j_k is conformal, hence $K[j_2 \circ f_0 \circ j_1] = K[f_0]$. By the uniqueness of the Teichmüller map, $j_2 \circ f_0 = f_0 \circ j_1$. Thus f_0 covers an admissible map $F_0: S_1 \to S_2$. F_0 is thus the unique extremal for the problem of minimizing the astigmatism among all maps homotopic to $id: S_1 \to S_2$.

We may assume that as a point set $S_1' = S_2'$ and f_0 is the identity map on S_1'. Then, by Teichmüller's theorem for compact surfaces, there is some $\omega \in Q_{S_1'} \setminus \{0\}$ so that

$$f_0: S_1' \to (S_1', \frac{\omega}{\|\omega\|}, \|\omega\|)$$

unless f_0 is conformal. The case where f_0 is conformal is trivial. We concentrate on the general case. Locally, ω projects to a quadratic differential on the punctured surface S_1. We must show that ω projects globally. If ω is not symmetric with respect to the involution j_1, then, using the above method, the Teichmüller deformation

$$S_1' \to (S_1', \frac{\omega \circ j_1}{\|\omega \circ j_1\|}, \|\omega\|)$$

also covers F_0. By the uniqueness part of Teichmüller's theorem, $\omega \circ j_1 = \omega$. ω thus projects to a non-zero quadratic differential Ω on the punctured surface S_1.

It is important to determine the nature of the possible singularities at punctures. Let ζ_0 be a puncture. Near $\zeta_1 \in p_1^{-1}(\zeta_0)$, assume ω has the normalized form $z^k dz^2$ for $k \geq 0$. k is even since ω is symmetric about ζ_1. $\zeta = p_1(z) = z^2$ hence locally

$$\Omega = \frac{1}{4\zeta} \zeta^{k/2} d\zeta^2.$$

If $k = 0$, Ω has a simple pole at ζ_0, otherwise Ω is holomorphic near ζ_0.

We have shown

<u>Lemma 1</u>: Let S_1 be a surface of type $(g,2n,0) \neq (0,2,0), (0,4,0)$

with $n \neq 0$ and assume S_2 is another Riemann surface structure on S_1. Then among all admissible maps $f\colon S_1 \to S_2$ homotopic to the identity, there is a Teichmüller deformation which is the unique extremal for minimizing $K[f]$.

We next extend Lemma 1 to surfaces of finite conformal type with an odd number $n > 1$ of punctures. Since the conformal group of the Riemann sphere $\hat{\mathbb{C}}$ is triply transitive, all surfaces of type $(0,3,0)$ are conformally equivalent and we may ignore type $(0,3,0)$. Let S_2 again denote a new conformal structure on the surface S_1 of type $(g,n,0) \neq (0,3,0)$ and let $f\colon S_1 \to S_2$ be an admissible map homotopic to the identity. Let S_1' be a 2-sheeted cover of S_1 with branch points of order 2 at the punctures $\zeta_1,\ldots,\zeta_{n-1}$. Further let S_2' be the corresponding cover of S_2. The puncture ζ_n has two preimages ζ_n', z_n' in S_1'. Thus S_1' may be viewed as a surface with an even number of punctures. We construct a 2-sheeted cover S_i'' of S_i' with branch points of order 2 over ζ_n' and z_n'. S_i'' is compact. f lifts to an admissible map $f_4\colon S_1'' \to S_2''$ which is homotopic to the identity. Since S_1'' has genus $4g + n - 3 > 1$, Teichmüller's theorem shows that

$$K[f_4] \geq K[F_4]$$

where $F_4\colon S_1'' \to S_2''$ is the Teichmüller deformation. The relevant parts of the proof of Lemma 1 may be repeated to show

<u>Lemma 2</u>: Let S_1 be a surface of type $(g,n,0) \neq (0,2,0),(0,4,0)$ and $n \neq 0,1$. Assume S_1 is hyperbolic. If S_2 is another Riemann surface structure on S_1 of type $(g,n,0)$, then, among all admiss-

ible maps $f: S_1 \to S_2$ homotopic to the identity, there is a Teichmüller deformation which is the unique map minimizing $K[f]$.

Assume now $n = 1$ and $g > 0$. We construct an unbranched 2-sheeted cover. The puncture lifts to two punctures and using Lemma 1 we obtain the existence and uniqueness of the Teichmüller deformation for once-punctured surfaces of positive genus.

One important case has been omitted. If S_1 has type $(0,4,0)$, we construct a 2-sheeted cover S_1' branched of order 2 at the punctures. S_1' is a torus, the points covering the punctures being distinguished. Thus we may use Lemma 1 to assert the existence and uniqueness of the Teichmüller deformation.

We summarize these results as

Theorem 1: Let S_1 be a Riemann surface of type $(g,n,0) \neq (0,1,0)$, $(0,2,0)$, $(1,0,0)$ and S_2 be another Riemann surface structure on S_1 of type $(g,n,0)$. Assume $id: S_1 \to S_2$ is not homotopic to a conformal map. Then there exists a Teichmüller deformation $F: S_1 \to S_2$ homotopic to the identity. Among all admissible maps $f: S_1 \to S_2$ homotopic to the identity, F is the unique map minimizing $K[f]$.

Theorem 1 is Teichmüller's theorem for surfaces of finite conformal type. We seek to generalize Theorem 1 to surfaces of finite topological type. This is done as in Lemma 1 by doubling the surface and then applying Theorem 1. Precisely we have

Theorem 2: (Teichmüller's theorem for surfaces of finite topological type) Let S_1 be a Riemann surface of type $(g,n,m) \neq (0,0,2), (0,1,1)$

whose universal cover is conformally the unit disc. Further, let S_2 *be another Riemann surface structure on* S_1 *with punctures and ideal boundary curves preserved. Then there exists a Teichmüller deformation* $F: S_1 \to S_2$ *homotopic to the identity. Among all admissible maps* $f: S_1 \to S_2$ *homotopic to the identity,* F *is the unique map minimizing* $K[f]$.

(1.6) The Fricke and Teichmüller spaces of a finite hyperbolic surface

Let S_0 be a finite hyperbolic surface and $\mathcal{Q}_{S_0}$ be the space of admissible quadratic differentials on S_0. $\mathcal{Q}_{S_0}$ is the space of meromorphic quadratic differentials ω on S_0 which satisfy

(i) $\omega = \omega^d \mid S_0$ where $\omega^d \in \mathcal{Q}_{S^d}$

(ii) ω is real on ∂S_0

(iii) ω has, at worst, a simple pole at each puncture $\zeta_0 \in S_0$.

As in the previous Chapter, $\|\omega\| = \iint_S |\omega| < \infty$ and the Teichmüller deformations of S_0 may be identified with the unit ball $\mathcal{Q}^1_{S_0}$ in $\mathcal{Q}_{S_0}$. As before, $0 \in \mathcal{Q}_{S_0}$ is associated to the trivial deformation.

Definition: If S_0 is a finite hyperbolic Riemann surface, the Teichmüller space, $T(S_0)$, of S_0 is the set of Teichmüller deformations of S_0 with the topology induced by the identification with $\mathcal{Q}^1_{S_0}$.

We shall need the next Lemma to justify some subsequent normalizations.

<u>Lemma</u>: Let G be a fuchsian group and $\gamma, \eta \in G$. Then γ and η share a common fixed point if and only if both lie in a cyclic subgroup of G.

<u>Proof</u>: Sufficiency is trivial. To show necessity, assume ∞ is the fixed point. If γ and η do not lie in a cyclic subgroup, simple computations of several distinct cases show that G is not discrete. We perform one of the computations and leave the remainder as exercises. Assume γ, and η are hyperbolic and have two common fixed points. If necessary by conjugating G, we may assume that the fixed points are 0 and ∞, i.e., $\gamma: z \mapsto \lambda_1 z$ and $\eta: z \mapsto \lambda_2 z$ with $\lambda_i \in \mathbb{R}^+ \setminus \{0\}$. Then $\gamma^k \eta^\ell: z \mapsto \lambda_1^k \lambda_2^\ell z$. Unless λ_1 and λ_2 are powers of some fixed λ, neither $\{\lambda_1^k \lambda_2^\ell\}$ nor G is discrete. ///

Let S_0 be a finite hyperbolic surface of type (g,n,m) whose double is hyperbolic. We wish to define the Fricke space $F_{g,n,m}$ of marked surfaces of type (g,n,m). A <u>marking</u> of S_0 is a set of homotopically non-trivial simple closed curves $\alpha_1,\ldots,\alpha_{2g},\beta_1,\ldots,\beta_n,$ $\delta_1,\ldots,\delta_m$ which generate $\pi_1 S_0$ and satisfy:

(M1) $$\#(\alpha_i,\alpha_j) = \begin{cases} \delta_{i+1,j} & \text{if } i \text{ is odd} \\ \delta_{i-1,j} & \text{if } i \text{ is even} \end{cases}$$

(M2) $$\#(\beta_i,\alpha_j) = \#(\beta_i,\beta_j) = \#(\beta_i,\delta_j) = 0$$

(M3) $$\#(\delta_i,\alpha_j) = \#(\delta_i,\delta_j) = 0.$$

Here $\#(\ ,\ )$ is the geometric intersection number, i.e., the minimal number of points of intersection as the curves are varied in their respective homotopy classes.

Ξ is the set of monomorphisms $\chi: \pi_1 S_0 \to \Gamma$ such that

(X1) Im χ is discrete

(X2) $\chi(\alpha_i)$ and $\chi(\delta_i)$ are hyperbolic

(X3) $\chi(\beta_i)$ is parabolic

with the following complicated normalization

(X4) if $g > 0$,

$\chi(\alpha_{2g})$ has 0 as a repelling fixed point

$\chi(\alpha_{2g})$ has ∞ as an attractive fixed point

$\chi(\alpha_{2g-1})$ has 1 as a fixed point

(X5) if $g = 0$ and $m > 0$, assume $\chi(\pi_1 S_0)$ acts discontinuously on $\mathbb{R}^- = \{\mathrm{Im}\, z = 0,\ \ \mathrm{Re}\, z < 0\}$

$\chi(\delta_m)$ has 0 as a repelling fixed point

$\chi(\delta_m)$ has ∞ as an attractive fixed point

and, if $m > 1$,

$\chi(\delta_{m-1})$ has 1 as an attractive fixed point

or, if $m = 1$,

$\chi(\beta_n)$ has 1 as fixed point

(X6) if $g = m = 0$, then $n > 2$ and

$\chi(\beta_n)$, $\chi(\beta_{n-1})$ and $\chi(\beta_{n-2})$ have 0 , 1 and ∞ respectively as their fixed points.

It is important to note that Ξ may be empty. This occurs only in genus 0 and is due either to an unfortunate choice of orientation of δ_m if $m > 0$ or an unfortunate ordering of the δ_m. This problem may be avoided if we assume that the marking has been chosen so that $\Xi \neq \emptyset$. Henceforth we make that assumption. Notice that if $m + n \neq 0$, $\pi_1 S_0$ is a free group on $2g + n + m - 1$ generators.

We next define a map $\mathfrak{F} = \mathfrak{F}_{g,n,m} : \Xi \to R^{6g-6+3m+2n}$. Let

$$\chi(\alpha_i): z \mapsto (a_i z + b_i)/(c_i z + d_i)$$

$$\chi(\beta_i): z \mapsto (a_i' z + b_i')/(c_i' z + d_i')$$

and

$$\chi(\delta_i): z \mapsto (a_i'' z + b_i'')/(c_i'' z + d_i'')$$

denote the generators of $\chi(\pi_1 S_0)$. Assume that the entries are chosen so that the traces are positive and the determinants are equal to one.

If $g > 0$, let

$$\mathfrak{F}(\chi) = (a_1, b_1, c_1, \ldots, a_{2g-2}, b_{2g-2}, c_{2g-2}, a_1', c_1', \ldots, a_n', c_n', a_1'', b_1'', c_1'', \ldots, a_m'', b_m'', c_m'').$$

If $g = 0$, $m > 1$, let

$$\mathfrak{F}(\chi) = (a_1', c_1', \ldots, a_n', c_n', a_1'', b_1'', c_1'', \ldots, a_{m-2}''. b_{m-2}'', c_{m-2}'').$$

If $g = 0$, $m = 1$, let

$$\mathfrak{F}(\chi) = (a_1', c_1', \ldots, a_{n-1}').$$

If $g = m = 0$ then $n > 3$ and we let

$$\mathfrak{F}(\chi) = (a_1', c_1', \ldots, a_{n-3}', c_{n-3}').$$

Note that we have not defined the map $\mathfrak{F}$ if $g = m = 0$ and $n = 3$.

In that case, there is no deformation possible.

The Fricke space $F_{g,n,m} = \mathcal{F}_{g,n,m}(\Xi)$. As in §1 of the previous Chapter, to show that $F_{g,n,m}$ parametrizes the conformal equivalence classes of marked surfaces of type (g,n,m), we must only show that $\chi(\pi_1 S_0)$ may be recovered from $\mathcal{F}(\chi)$. This is done exactly as in the previous Chapter using the relation

$$\{\prod_{i \text{ odd}} [\alpha_i, \alpha_{i+1}]\} \beta_1 \ldots \beta_n \delta_1 \ldots \delta_m = 1.$$

Since $\dim_{\mathbb{R}} Q_{S_0} = 6g - 6 + 2n + 3m$, the arguments of §4 of the previous Chapter may be repeated to show

Theorem: If S_0 is a marked hyperbolic surface of type (g,n,m) with hyperbolic double, then there is a canonical homeomorphism of $T(S_0)$ with $F_{g,n,m}$.

Aagin, as in Chapter I, $T(S_0)$ carries a basepoint-free natural metric and we may write $T_{g,n,m}$ instead of $T(S_0)$. $T_{g,n,m}$ is called the Teichmüller space of type (g,n,m).

§2 The Teichmüller modular group and the Riemann space

We have defined the Fricke and Teichmüller spaces for Riemann surfaces S of finite type whose doubles are hyperbolic. In this section we show that these spaces admit natural automorphisms. The group of automorphisms is the Teichmüller modular group, Mod. Mod acts discontinuously on the Teichmüller space. For compact surfaces S without boundary, $T(S)$ is biholomorphically equivalent to a bounded domain in $\mathbb{C}^{3g-3}$ for $g > 1$. In that case,

Royden has shown that Mod is the full group of holomorphic automorphisms of $T(S)$. Since we are only concerned with the real analytic structure of $T(S)$ we shall omit the (highly non-trivial) proofs of these last results. $R(S) = T(S)/\text{Mod}$ is called the Riemann space of S and parametrizes the conformal equivalence classes of Riemann surfaces of the same type as S. $R(S)$ may be given the structure of a normal complex space if S is of finite conformal type.

(2.1) <u>The action on</u> $T(S_0)$ <u>induced by diffeomorphisms of</u> S_0

Let S_0 be a marked Riemann surface of type (g,n,m) whose double is hyperbolic. We denote by $\text{Diff}_+ S_0$ the group of diffeomorphisms $f: S_0 \to S_0$ satisfying:

(Diff 1) f is orientation-preserving

(Diff 2) f extends to a diffeomorphism of S_0^d

(Diff 3) $K[f] < \infty$.

(Diff 2) and (Diff 3) imply that f maps punctures to punctures. Notice that $\text{Diff}_+ S_0$ depends only on the differentiable structure, not the conformal structure, of S as a punctured surface with boundary. We may therefore suppress the dependence on S_0 and write $\text{Diff}_+ (g,n,m)$ or simply Diff_+. If $f \in \text{Diff}_+$ then f induces a map

$$(f): \pi_1(S_0,\zeta_0) \to \pi_1(S_0,f(\zeta_0))$$
$$\{\alpha\}_{\zeta_0} \mapsto \{f \circ \alpha\}_{f(\zeta_0)}$$

where $\{\ \}_\zeta$ denotes the homotopy class of closed curves based at ζ. $f \circ \alpha$ is freely homotopic to a closed curve based at ζ_0, hence f induces an automorphism

$$\chi_{f,\beta}: \pi_1 S_0 \to \pi_1 S_0$$
$$\{\alpha\}_{\zeta_0} \mapsto \{\beta f(\alpha)\beta^{-1}\}_{\zeta_0}$$

where β is some fixed curve from $f(\zeta_0)$ to ζ_0. $\chi_{f,\beta}$ depends only on the homotopy class of β (with fixed endpoints). $\chi_{f,\beta}$ is an inner automorphism if and only if there is a curve β_1 from ζ_0 to $f(\zeta_0)$ so that, for all $\{\alpha\}_{\zeta_0} \in \pi_1 S_0$,

$$\{f(\alpha)\}_{f(\zeta_0)} = \{\beta_1\}\{\alpha\}_{\zeta_0}\{\beta_1\}^{-1}.$$

Here $\{\beta_1\}$ denotes the homotopy class with fixed endpoints.

The automorphism $\chi_{f,\beta}$ of $\pi_1 S_0$ is called a <u>change of marking</u>.

Each $f \in \mathrm{Diff}_+$ lifts to a diffeomorphism f of U. By (Diff 3) and the Ahlfors-Bers theorem (see the previous Chapter), f extends to a homeomorphism of $U \cup \hat{\mathbb{R}}$, where $\hat{\mathbb{R}} = \mathbb{R} \cup \{\infty\}$. We may therefore normalize f so that 0, 1 and ∞ are fixed. f conjugates the normalized group G of deck transformations for the covering $\pi: U \to S_0$ into a fuchsian group G_f. G_f is the normalized group of deck transformations for the universal covering $\pi': U \to S_0$ but here S_0 is marked by the image under f of the marking on S_0. This defines a new point $\chi_{f,S_0} \in \Xi_{g,n,m}$. Using the homeomorphisms between $\Xi_{g,n,m}$, $F_{g,n,m}$ and $T_{g,n,m}$, we have defined an action of Diff_+ on the latter two spaces. We denote by f^* the map on $F_{g,n,m}$ or $T_{g,n,m}$ induced by $f \in \mathrm{Diff}_+$.

Henceforth, whenever the meaning is clear, we shall denote points in $F_{g,n,m}$ and $T_{g,n,m}$ by the marked surfaces that they represent.

A special role is played by those diffeomorphisms homotopic or diffeotopic to the identity. We let $\mathrm{Diff}_0 = \mathrm{Diff}_0(g,n,m)$ denote the set of those diffeomorphisms $f \in \mathrm{Diff}_+$ which are homotopic to the identity.

<u>Lemma 1</u>: If $f \in \mathrm{Diff}_0$ then $f^* = \mathrm{id}$.

<u>Proof</u>: Let S correspond to the monomorphism $\chi \in \Xi_{g,n,m}$. S and $f(S)$ are the same Riemann surface and therefore have groups G and G' of deck transformations which are conjugate in Aut U. $\chi_{f,\beta}$ gives the change of generators up to an inner automorphism i_β of $\pi_1 S_0$. Thus

$$\chi_{f,S}(\alpha) = \eta(\chi \circ i_\beta \circ \chi_{f,\beta}(\alpha))\eta^{-1}$$

where η reestablishes the normalization. Since f is homotopic to the identity, $\chi_{f,S}$ is conjugation by a fixed curve. Thus

$$\chi_{f,\beta}(\alpha) = \eta\gamma\chi(\alpha)\gamma^{-1}\eta^{-1}$$

for some $\gamma \in G'$. The normalization gives $\eta\gamma = \mathrm{id}$. Therefore $\chi_{f,S} = \chi$ and $f^*(S) = S$ hence $f^* = \mathrm{id}$. ///

<u>Lemma 2</u>: If $f \in \mathrm{Diff}_+$ then f^* is a homeomorphism of $F_{g,n,m}$ hence of $T_{g,n,m}$.

<u>Proof</u>: As in the previous lemma, χ_{f,S_0} is determined by a conjuga-

tion and $\chi_{f,\beta}$. Since $\pi_1 S_0$ is finitely generated $\chi_{f,\beta}([\alpha])$ is a finite word in the finite number of generators of $\pi_1 S_0$ and the inverses of those generators. Let $C = \{\alpha_1, \ldots, \alpha_k\}$ be the given marking of S_0. The matrix entries of $\chi \circ \chi_{f,\beta}([\alpha_i])$ are clearly polynomials in the matrix entries of the $\chi([\alpha_i])$ and the $\chi([\alpha_i]^{-1})$ hence are continuous. The renormalization is also continuous hence f^* is continuous. $(f^{-1})^* = (f^*)^{-1}$ which shows that f^* is a homeomorphism. Since $T_{g,n,m}$ is homeomorphic to $F_{g,n,m}$, the last conclusion follows. ///

By abuse of language we say that f acts on $F_{g,n,m}$ or $T_{g,n,m}$ by <u>change of marking</u>.

<u>Theorem 1</u>: <u>Let</u> $S \in F_{g,n,m}$ <u>have hyperbolic double</u>. <u>If</u> $f \in \mathrm{Diff}_+(g,n,m)$ <u>then</u> $f^*(S) = S$ <u>if and only if</u> f <u>is homotopic to a conformal self-map of</u> S.

<u>Proof</u>: If h is a conformal self-map of S, then h is covered by some $\gamma \in \mathrm{Aut}\, U$. $G' = \eta\gamma G\gamma^{-1}\eta^{-1}$ where η reestablishes the normalization. Clearly $\eta = \gamma^{-1}$ and $G' = G$. Each generator γ_k' for G', as a marked group, is given by $\gamma_k' = \eta\gamma\gamma_k(\eta\gamma)^{-1}$ where γ_k is the corresponding generator for G. Thus $\gamma_k' = \gamma_k$ and $h^*(S) = S$. If f is homotopic to the conformal map h, then $f \circ h^{-1} \in \mathrm{Diff}_0(g,n,m)$ and by Lemma 1, $f^*(S) = S$.

To prove the converse, let S_1 be S with the conformal structure of S pulled back by f^{-1}. Let h denote the map from S_1 to S which as a point map is f. h is then holomorphic and f factors as

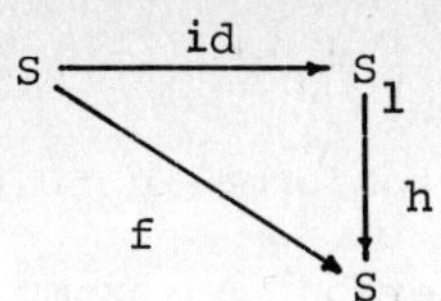

The Teichmüller distance between $f^*(S)$ and S is zero, hence $\mathrm{id}: S \to S_1$ is homotopic to a conformal map. ///

The (differentiable) <u>mapping class group</u> $M = M_{g,n,m}$ is $\mathrm{Diff}_+/\mathrm{Diff}_0$. By the previous theorem, Diff_0 lies in the subgroup $I = I_{g,n,m}$ of Diff_+ acting trivially on $T_{g,n,m}$. The <u>Teichmüller modular group</u> $\mathrm{Mod} = \mathrm{Mod}(g,n,m)$ is Diff_+/I.

If S is of finite topological type and either S or S^d is not hyperbolic, then $(g,n,m) = (0,0,0)$, $(0,0,1)$, $(0,0,2)$, $(0,1,0)$, $(0,1,1)$, $(0,2,0)$ or $(1,0,0)$. These are called the <u>excluded types</u>.

<u>Theorem 2</u>: <u>If</u> $g \in \mathrm{Mod}(g,n,m)$ <u>then</u> g <u>is an isometry of</u> $T_{g,n,m}$ <u>in the Teichmüller metric</u>.

<u>Proof</u>: Let $S_1, S_2 \in T_{g,n,m}$ and $d = d_T(S_1,S_2)$. If $h: S_1 \to S_2$ is the Teichmüller mapping realing d, then h is also the extremal admissible map if we consistently change the markings on S_1 and S_2. But this is precisely the action of g. Thus $d_T(g(S_1),g(S_2)) = d$. ///

<u>Corollary</u>: If (g,n,m) is not an excluded type, then $\mathrm{Mod}(g,n,m) = M_{g,n,m}$ unless the following condition holds: there exists $f \in \mathrm{Diff}_+ \setminus \mathrm{Diff}_0$ so that for all Riemann surface structures S of type (g,n,m), $f: S \to S$ is homotopic to a conformal map.

The non-excluded types (g,n,m) for which $\mathrm{Mod}(g,n,m) \neq M_{g,n,m}$ is quite a short list. For conformally finite surfaces, the list has been given by David Patterson. An example is type $(2,0,0)$. Every compact unbordered surface of genus 2 may be represented as a two-sheeted cover of $\hat{\mathbb{C}}$ branched over six points. The conformal self-map is the sheet interchange. For all other types $(g,0,0)$ with $g > 2$, $\mathrm{Mod}(g,n,m) = M_{g,n,m}$.

__Definition__: Let (g,n,m) be a non-excluded type. Then the __Riemann__ space of type (g,n,m) is

$$R = R(S_0) = R_{g,n,m} = T_{g,n,m}/\mathrm{Mod}(g,n,m).$$

(2.2) __Discontinuity of the action of__ $\mathrm{Mod}(g,n,m)$ __on__ $T_{g,n,m}$

In this section we assume (g,n,m) is not an excluded type. We wish to show that $\mathrm{Mod}(g,n,m)$ acts discontinuously on $T_{g,n,m}$. Our proof, while not necessarily the most direct, gives the discontinuity of the action as a consequence of the discreteness of the length spectrum for surfaces S of finite non-excluded conformal type. The length spectrum of an unbordered surface S itself is of much contemporary interest since, via the Selberg trace formula, it may be related to the eigenvalues of the Laplacian for S. Wolpert has recently shown that, save on a real analytic sublocus of $T(S)$, the length spectrum of S specifies S up to conformal or anti-conformal equivalence. Vigneras has recently announced the existence of non-equivalent surfaces with the same length spectrum.

<u>Definition</u>: Let S be a surface of non-excluded type (g,n,m) and $C = (\alpha_i)$ be a list of representatives of all elements of $\pi_1 S$. The <u>ordered length spectrum</u> $\mathcal{L}_o(S,C)$ is the sequence $(\ell\{\alpha_i\})$ where

(i) $\ell\{\alpha_i\}$ is the hyperbolic length of the geodesic in the free homotopy class of α_i, or

(ii) $\ell\{\alpha_i\} = 0$ if $\{\alpha_i\}$ contains no geodesics.

The <u>length spectrum</u> of S is the set

$$\mathcal{L}(S) = \{\ell\{\alpha_i\} \mid \alpha_i \in \pi_1 S\}.$$

<u>Lemma 1</u>: Suppose G is a fuchsian group with $G \ni \gamma: z \mapsto z + 1$.

If $H = \{\operatorname{Im} z > 1\}$ and γ is primitive in G, then, for all $\eta \in G$, either

(i) $\eta(H) \cap H = \emptyset$, or

(ii) $\eta = \gamma^n$ for some $n \in \mathbb{Z}$.

<u>Proof</u>: Assume $\eta: z \mapsto (az+b)/(cz+d)$, $ad - bc = 1$ and $\eta \neq \gamma^n$. Define inductively

$$\eta_{k+1} = \eta_k \circ \gamma \circ \eta_k^{-1}: z \mapsto (a_k z+b_k)/(c_k z+d_k)$$

with $\eta_1 = \eta$ and $a_k d_k - b_k c_k = 1$. Then

$$a_{k+1} = 1 - a_k c_k$$

$$b_{k+1} = a_k^2$$

$$c_{k+1} = -c_k^2.$$

If $|c| < 1$, then $c_k = -c^{2^k} \to 0$. But then a_k, b_k and d_k converge to 1, $\eta_k \to \gamma$ and G is not discrete. This contradiction establishes that $|c| \geq 1$.

Let $z \in H$, then $\operatorname{Im} \eta(z) = (\operatorname{Im} z)/|cz+d|^2$. $H' = \eta(\partial H)$ is a circle tangent to $\mathbb{R}$ at a/c if η is not a power of γ.

$$\begin{aligned}\operatorname{diam} H' &= \sup_{x \in \mathbb{R}} \left| \frac{a(x+i) + b}{c(x+i) + d} - \frac{a}{c} \right| \\ &= \sup_{x \in \mathbb{R}} \frac{1}{|c|^2} \left| \frac{1}{x + i + (d/c)} \right| \\ &= |c|^{-2} \\ &\leq 1\end{aligned}$$

which immediately implies the desired result. ///

<u>Definition</u>: In the notation of the previous lemma, H is called a <u>horocycle</u> at the fixed point of γ.

The invariant formulation of Lemma 1 is that to each cyclic parabolic subgroup P of G, there is associated a horocycle H with the following property. If GH denotes the orbit of H then GH/G is conformally equivalent to H/P which is a punctured disc.

<u>Lemma 2</u>: Let S be a hyperbolic Riemann surface. Further suppose G is a fuchsian group with $U/G = S$ and $\gamma \in G$ covers a class $\{\alpha\} \in \pi_1 S$. Then the following are equivalent:

(i) γ is parabolic

(ii) $\{\alpha\}$ is homotopic to a power of a simple loop around a puncture.

<u>Proof</u>: Corollary 1.1.1 shows that (ii) implies (i). We show

that (i) implies (ii). By conjugating G we may assume $\gamma: z \mapsto z+1$ and that γ is primary. If $\pi: U \to S$ is the projection map, α is homotopic to $\pi(\{i+t \mid t \in [0,1]\})$. As remarked above the latter is a simple loop around a puncture of S. ///

<u>Lemma 3</u>: Let $S = U/G$ be a hyperbolic Riemann surface. Assume $\{\alpha\} \in \pi_1 S$ and $G \ni \gamma: z \mapsto (az+b)/(cz+d)$ covers $\{\alpha\}$ with $ad - bc = 1$. Then

[2.1] $$\ell\{\alpha\} = 2 \left| \log |\epsilon| \right|$$

where ϵ is an eigenvalue of

$$\gamma^* = \begin{bmatrix} a & b \\ c & d \end{bmatrix}$$

and

[2.2] $$\mathrm{tr}^2 \gamma^* = 4 \cosh^2 \frac{\ell\{\alpha\}}{2}.$$

<u>Proof</u>: In §1.1 we showed that parabolic elements cover homotopy classes with no geodesics. Thus [2.1] and [2.2] are valid if $\epsilon = \pm 1$. If not, we may assume that γ is hyperbolic. Further, by conjugating G, we may assume $\gamma: z \mapsto \lambda z$ with $\lambda > 1$. The length of the geodesic $[\alpha]$ in $\{\alpha\}$ is

$$\ell\{\alpha\} = \int_{\lambda}^{i\lambda} y^{-1} dy = \log \lambda.$$

$\lambda = \epsilon^2$. [2.2] follows from [2.1] and the invariance of the trace under conjugation. ///

<u>Theorem 1</u>: <u>Let</u> S <u>be a hyperbolic Riemann surface of type</u> $(g,n,0)$.

<u>Then the sequence</u> (ℓ_i) <u>of lengths of closed geodesics on</u> S <u>is discrete</u>.

<u>Proof</u>: Let (α_i) be a sequence of distinct closed geodesics on S and assume that $\ell\{\alpha_i\} < M < \infty$. Choose $p \in S$ and $p_i \in \alpha_i$ so that $d(p,\alpha_i) = d(p,p_i)$ where d denotes hyperbolic distance. Let $\pi: U \to S$ be the universal covering with cover group G. If $\tilde{p} \in \pi^{-1}(p)$, there exists $\tilde{p}_i \in \pi^{-1}(p_i)$ so that $d(\tilde{p},\tilde{p}_i) = d(p,p_i)$. If $\gamma \in G$ covers α_i and $\tilde{p}_i$ lies on the geodesic in U fixed by γ_i then

$$d(\tilde{p},\gamma_i(\tilde{p})) \leq 2d(p,p_i) + \ell(\alpha_i).$$

Since G acts discontinuously on U, $d(\tilde{p},\gamma_i(\tilde{p})) \to \infty$. It follows that p_i diverges on S. We may therefore choose a subsequence (p_i') so that p_i' converges to a single puncture on S. Then $\tilde{p}_i$ may be assumed to converge to a parabolic fixed point. By conjugation we may assume that the corresponding parabolic transformation is $z \mapsto z + 1$. Thus $\text{Im}\,\tilde{p}_i \to \infty$ and, by Lemma 1, $\text{Im}\,\gamma_i(\tilde{p}_i) < 1$. Thus $\ell(\alpha_i) = d(\tilde{p}_i,\gamma_i(\tilde{p}_i)) \to \infty$ which is a contradiction. ///

We have the immediate

<u>Corollary 1</u>: For any list C of $\pi_1 S$, $\mathcal{L}_o(S,C)$ and $\mathcal{L}(S)$ are discrete if S has finite non-excluded conformal type.

If S has type (g,n,m) which is not excluded, then the infinite Nielsen extension S_∞ of S has type (g,n+m,0) which is also not excluded. If α is a closed curve on S, we may compute the length $\ell\{\alpha\}$ (respectively $\ell\{\alpha\}_\infty$) of the geodesic in the class of α in S (respectively in S_∞). By Theorem (1.1), $\ell\{\alpha\}_\infty < \ell\{\alpha\}$. Thus for any list C of $\pi_1 S$, the ordered length spectrum $\mathcal{L}_o(S,C)$ is termwise larger than the discrete sequence $\mathcal{L}_o(S_\infty,C)$. We immediately obtain

Corollary 2: If S has type (g,n,m) which is not excluded, then $\mathcal{L}(S)$ and $\mathcal{L}_o(S,C)$ are discrete.

Definition: Let S and S' be Riemann surfaces marked by curves $\{\alpha_1,\ldots,\alpha_N\}$ and $\{\alpha'_1,\ldots,\alpha'_N\}$ respectively. A diffeomorphism $f\colon S \to S'$ is said to preserve marking if, for all i, $f \circ \alpha_i$ is homotopic to α'_i.

Our next theorem shows that, on surfaces whose types are not excluded, the lengths of a finite number of geodesics specifies a surface in $F_{g,n,m}$, unique up to an anti-conformal map.

Theorem 2: Let S and S' be marked Riemann surfaces of non-excluded type (g,n,m) with markings $\{\alpha_1,\ldots,\alpha_N\}$ and $\{\alpha'_1,\ldots,\alpha'_N\}$. Further let $f\colon S \to S'$ be an orientation preserving and marking preserving diffeomorphism. Then there exists a finite set $\{\beta_1,\ldots,\beta_M\}$ of closed curves on S so that $\ell\{f \circ \beta_i\} = \ell\{\beta_i\}$ for all i implies f is homotopic to a conformal map. Moreover the curves β_i depend only on the type of S.

Proof: It suffices to show that S and S' determine the same point in $F_{g,n,m}$. Equivalently, it suffices to show that, if $S' = U/G'$ with G' normalized, then the coefficients of the Möbius transformations generating G' may be recovered from $\ell\{\beta_i\}$ and the fact that f preserves orientation. Let G be the normalized cover group for the projection $\pi: U \to S$.

If $(g,n,m) \neq (0,n,0)$, there is a generator of $\pi_1 S$ which is not retractible into a puncture; relabel so that the given generator is α_1. We normalize G and G' so that

(i) α_1 (respectively, $f \circ \alpha_1$) is covered by $\gamma_1 : z \mapsto \mu^2 z$ with $\mu > 1$ (respectively, $\gamma_1' : z \mapsto \lambda^2 z$ with $\lambda > 1$)

(ii) α_2 (respectively, $f \circ \alpha_2$) is covered by γ_2 (respectively, γ_2') and $\gamma_2(\infty) = \gamma_2'(\infty) = \pm 1$.

The choice called for in (ii) is possible because of Lemma (1.5).

Notice that $\gamma_2(\infty)$ and $\gamma_2'(\infty)$ have the same sign because is orientation preserving. Let $\gamma_i' : z \mapsto (a_i z + b_i)/(c_i z + d_i)$ cover α_i' with $a_i d_i - b_i c_i = 1$. If necessary, by multiplying all coefficients by -1, we may assume that $a_i + d_i \geq 0$. Let $\rho_i = \mathrm{tr}^2\, \gamma_i'$, $\tau_i = \mathrm{tr}^2(\gamma_1' \gamma_i')$ and $\sigma_i = \mathrm{tr}^2((\gamma_1')^{-1} \gamma_i')$ for $i = 2, \ldots, M$. Then

[2.3]
$$\lambda a_i + \lambda^{-1} d_i = \pm \sqrt{\tau_i}$$
$$\lambda^{-1} a_i + \lambda d_i = \pm \sqrt{\sigma_i}$$

Addition gives

[2.4]
$$(\lambda + \lambda^{-1})\ (a_i + d_i) = \pm \sqrt{\sigma_i} \pm \sqrt{\tau_i}$$

and squaring, we obtain

[2.5]
$$(\lambda + \lambda^{-1})^2\ \rho_i = \sigma_i + \tau_i \pm 2\sqrt{\sigma_i \tau_i}.$$

As noted above, ρ_i, σ_i and τ_i are known, hence [2.5] gives the choice of sign of $\sqrt{\sigma_i \tau_i}$. Since $a_i + d_i > 0$, [2.4] then determines the choices of sign in [2.3] and a_i and d_i are known.

$c_2 = \pm a_2$ depending only on the known sign of $\gamma_2(\infty)$. b_2 is then given by the determinant condition. To find c_i and d_i for $i > 2$ we proceed as above using $\gamma_2'\gamma_i'$ and $\gamma_2'(\gamma_i')^{-1}$. G' is then completely determined.

If $(g,n,m) = (0,n,0)$, replace α_1 and α_2 by $\alpha_1\alpha_2$ and α_2^{-1}. $\alpha_1\alpha_2$ is not retractible into a puncture and must be covered by a hyperbolic transformation when $n > 3$. The above argument may then be used to complete the proof. ///

L. Keen has shown that the lengths of $6g - 6 + 2n + 3m$ closed curves suffice to determine the structure of S.

Our next two results are standard facts in Riemann surface theory. Associated to the Poincaré metric ρ in U is an area element $dA_\rho = y^{-2}dxdy$. dA_ρ is Aut U invariant, hence projects to any hyperbolic surface. $A_\rho(S) = \iint_S dA_\rho$ is the <u>hyperbolic area</u> of S.

<u>Lemma 4</u>: If S is a conformally finite hyperbolic surface then $A_\rho(S) < \infty$.

<u>Proof</u>: Let $\zeta_1,\ldots,\zeta_k$ be a list of the punctures on $S = U/G$ and N_i be a small disc with center ζ_i. Since $S \setminus \bigcup_i N_i$ is compact, $A_\rho(S \setminus \bigcup_i N_i) < \infty$. If α_i is a small loop about ζ_i, we may assume α_i is covered by $\gamma: z \mapsto z + 1$. Then for N_i sufficiently small, $N_i \subset \pi(B)$ where $B = \{z \in U \mid |\text{Re } z| \leq 1/2, \text{ Im } z > 1\}$. But $A_\rho(N_i) \leq A_\rho(B) = 1$ which shows that $A_\rho(S) < \infty$. ///

It is known that, for a conformally finite hyperbolic surface of type $(g,n,0)$, $A_\rho(S) = 2\pi(2g+n-2)$. Two reasonably elementary proofs are known. The first proof uses the fact that S has curvature = -1, the Gauss-Bonnet theorem and exhaustion. The second is more elementary, but more complicated. One computes the hyperbolic area of a triangle, then uses the invariance of the Euler characteristic to compute $A(S)$ (c.f., Macbeath [6]).

Our next result is a basic fact in the theory of Riemann surfaces. The length of the proof precludes its inclusion here. Siegel's proof [10, p. 46] is quite accessible.

Definition: Let G be a fuchsian group. A fundamental set P for G is an open set $P \subset U$ so that $A_\rho(\partial P) = 0$ and which is a maximal open set on which $\pi: U \to U/G$ is injective.

Lemma 5: Let G be a fuchsian group with fundamental set P, then $A_\rho(P) \geq \frac{\pi}{21}$.

Lemma 6: Let G be a fuchsian group and $N(G)$ be the normalizer of G in Aut U, then $N(G)$ is fuchsian.

Proof: If not, there exists a sequence $\gamma_n \to \text{id}$ in $N(G)$. If $\gamma \in G$, then $G \ni \gamma_n \cdot \gamma \cdot \gamma_n^{-1} = \eta_n \to \gamma$. This contradicts the fact that G is fuchsian. ///

Theorem 3: Let S be a Riemann surface of non-excluded type (g,n,m) and Aut S be the holomorphic automorphism group of S. Then Aut S is a finite group.

Proof: Let $\pi: U \to U/G = S$ be the universal cover map. Each $f \in$ Aut S is covered by a Möbius transformation $\gamma \in$ Aut U which normalizes G. Conversely each $\gamma \in N(G)$ covers some $f \in$ Aut S. Thus Aut $S \simeq N(G)/G$.

Since $N(G)$ is fuchsian it has a fundamental set P_N with $A_\rho(P_N) \geq \frac{\pi}{21}$.

If S has finite conformal type, let $\{\gamma_i\}$ be the coset representatives in $N(G)/G$. $\gamma_i(P_N) \cap \gamma_j(P_N) = \emptyset$ for $i \neq j$ since $\pi_N: U \to U/N(G)$ is injective on P_N. $P = \cup \gamma_i(P_N)$ lies in a fundamental set for G. Since $A_\rho(S) < \infty$ so is $A_\rho(P)$.

But $A_\rho(P) = \sum_i A_\rho(P_N) \leq A_\rho(S) < \infty$, thus both the index set $\{i\}$ and $\operatorname{Aut} S$ are finite.

If S is not conformally finite, each conformal automorphism of S extends to a conformal automorphism of the double S^d of S. The above argument then shows that $\operatorname{Aut} S$ is finite. ///

Theorem 4: If S has non-excluded type (g,n,m), then $M_{g,n,m}$ acts discontinuously on $T(S)$.

Proof: Suppose not. Then there exists $g_k \in M_{g,n,m}$ so that $S_k = g_k(S_1) \to S'$. If f_k is a diffeomorphism inducing g_k then f_k determines a map

$$f_k^*: \mathcal{L}_o(S_1) \to \mathcal{L}_o(S_k)$$
$$: \ell\{a\} \mapsto \ell\{f_k \circ \alpha\}$$

$\ell\{f_k \circ \alpha\}$ is independent of the choice of f_k inducing g_k.

Since $T_{g,n,m}$ and $F_{g,n,m}$ are homeomorphic, the corresponding normalized cover groups converge elementwise. Further $\mathcal{L}_o(S_k) \to \mathcal{L}_o(S')$ as elements in $(\mathbb{R}^+)^\infty$. Since $\mathcal{L}(S_k)$ and $\mathcal{L}(S')$ are discrete, $f_k^*(\alpha)$ is eventually constant for each α. Thus for any finite collection $\{\beta_1,\ldots,\beta_N\}$ of closed geodesics and sufficiently large k,

$$f_{k_1}^* \circ (f_k^*)^{-1}(\ell(\beta_j)) = \ell(\beta_j)$$

for $j = 1,\ldots,N$ and all $k_1 > k$. By Theorem 2, there is a conformal map $S_{k_1} \to S_k$ which preserves markings. As points in the Teichmüller space $S_{k_1} = S_k$, for all $k_1 > k$. By the previous theorem, only finitely many $g_k \in M_{g,n,m}$ may be induced by a conformal automorphism of S_k and the sequence (g_k) is finite. ///

Among the immediate consequences of the theorem (or its proof) are

Corollary 1: Mod acts discontinuously on $T(S)$.

Corollary 2: $\text{Aut } S \simeq \text{Stab}_M S = \{g \in M_{g,n,m} \mid g(S) = S\}$.

Corollary 3: $R(S)$ is a Hausdorff space.

§3 Augmented Moduli spaces

Moduli spaces are parameter spaces for deformation of a given analytic or geometric object. Both the Teichmüller space and the Riemann space are moduli spaces for Riemann surfaces. In the Teichmüller space, there are globally defined coordinates given by either quadratic differentials - the Teichmüller coordinates - or matrix entries - the Fricke coordinates. In Section (3.2), we shall introduce another set of coordinates which are better suited to the study of the degeneration of structure of a given Riemann surface.

Mumford and Mayer using algebraic geometric techniques first showed that the space of conformal equivalence classes of compact Riemann surfaces of fixed genus has a natural compactification. The ideal points which are added represent Riemann surfaces on which a finite number of simple loops have been contracted to points (see Figure 1). This compactification of $R(S)$ is denoted $\hat{R}(S)$ and is called the augmented Riemann space. The ideal boundary of $\hat{R}$ corresponds to ideal boundary points which may be added to any

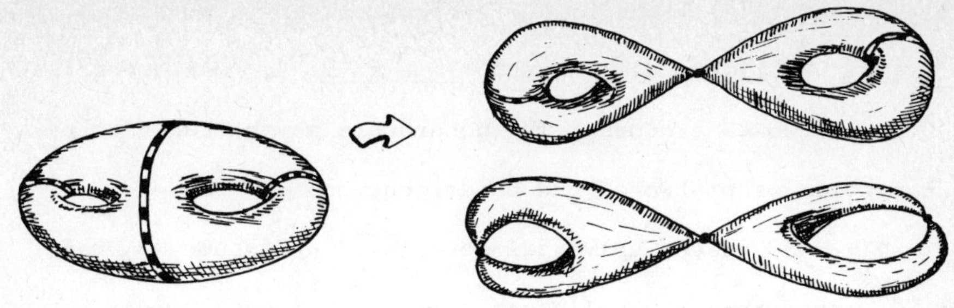

Figure 1

ramified cover of $R(S)$. In particular we obtain a completion $\hat{T}(S)$ of the Teichmüller space $T(S)$. Mod acts on $\hat{T}(S)$ and $\hat{R}(S) = \hat{T}(S)/\text{Mod}$.

First we must do some tailoring.

(3.1) Pants and their conformal structures

It is a classical and simple consequence of the classification of surfaces that each orientable surface of finite non-excluded type may be written as a finite connected sum of triply connected domains. The gluing occurs along one or more of the border curves. In this section, we shall describe the conformal structure of such triply connected domains, or, in our terminology, pants. In the following section, we shall describe the conformal gluing process. This allows us to parametrize the Teichmüller space by the lengths of the ideal border curves of the triply connected domains and certain gluing angles.

Following Thurston, we define pants (in French, pantalon) to be a bordered surface S, whose interior is homeomorphic to the triply connected domain $\{z \mid |z| < 4, |z-2| > 1, |z+2| > 1\}$. A pants, one or more of whose border components C is a puncture, is called tight (about C). The border components carry the suggestive names of waist W, right cuff R and left cuff L. One notes that $\pi_1 S$ is free on two generators. The generators may be taken to be the homotopy classes of two of the border curves.

Let $\pi:\Delta \to S$ be the universal covering. We assume $\zeta_o \in S$ and from ζ_o draw the hyperbolic geodesic C_i to each border component. $S\setminus(\cup C_i)$ is a simply connected Riemann surface and therefore lifts to a simply connected domain D. D is unique once we choose a point $\zeta \in S\setminus(\cup C_i)$ and a point $z \in \pi^{-1}(\zeta) \cap D$. (c.f., Figure 2).

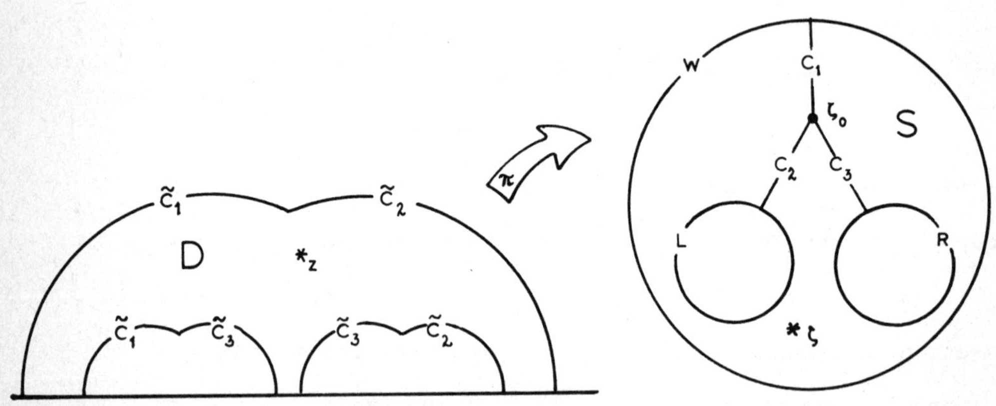

Figure 2

Theorem: If S is a pants, then the conformal structure of S is determined up to conformal or anti-conformal equivalence by $\ell\{W\}$, $\ell\{R\}$, and $\ell\{L\}$.

Proof: First orient the curves W, R, and L. We normalize the group G of deck transformations for the covering $\pi:U \to S$ so that $\gamma_1: z \mapsto \lambda^2 z$ with $\lambda > 1$, covers W, $\gamma_2: z \mapsto \frac{az+b}{cz+d}$, $\gamma_2(\infty) = \pm 1$, covers R and $\gamma_3 = \gamma_2\gamma_1$ covers L. Assume $\gamma_2(\infty) = 1$. The associated matrices are

$$\gamma_1^* = \begin{bmatrix} \lambda & 0 \\ & & -1 \\ 0 & \lambda \end{bmatrix}, \quad \gamma_2^* = \begin{bmatrix} a_2 & b_2 \\ c_2 & d_2 \end{bmatrix} \quad \text{and} \quad \gamma_3^* = \begin{bmatrix} a_3 & b_3 \\ c_3 & d_3 \end{bmatrix}.$$

By Lemma (2.2.3) $\text{tr}\ \gamma_i^*$ is known up to sign from $\ell\{W\}$, $\ell\{R\}$ and $\ell\{L\}$. Since γ_1 and γ_2 are free generators of G we may assume $K_2 = \text{tr}\ \gamma_2^* > 0$. For fixed choice of sign of $K_3 = \text{tr}\ \gamma_3^*$ there is a unique solution to the system of equations defined by

$$\gamma_1^*\gamma_2^* = \gamma_3^*$$

$$\text{tr}\ \gamma_i^* = K_i \qquad i = 1,2$$

$$a_2 = c_2 \quad .$$

Hence G and S are uniquely determined. Reversing the sign of K_3 multiplies γ_2^* by $-\text{id}$, hence does not change G.

Under the assumption that $\gamma_2(\infty) = -1$, the cover group G is as above except conjugated by reflection in the imaginary axis. Conjugation by this reflection exactly covers an anticonformal equivalence. ///

We shall later need the following corollary; first we have

<u>Definition</u>: If S is a bordered Riemann surface of non-excluded type then the <u>Nielsen Kernel</u> S^K of S is the bordered (connected) subsurface of S so that:

(i) S^K is a deformation retract of S

(ii) If $\alpha \subset S$ is a geodesic simple loop retractable into the border of S, then $\alpha \subset \partial S^K$.

Notice that S^K is the unique surface whose Nielsen extension is S.

Corollary 1: Let S_1 be a pants. If α is a closed curve on S_1 let $\ell_I(\alpha)$ denote the intrinsic length of α and

$$\ell_I\{\alpha\} = \inf\{\ell_I(\beta) \mid \beta \in [\alpha]\} .$$

S_1 is specified up to conformal or anti-conformal equivalence by $\ell_I\{W\}$, $\ell_I\{R\}$ and $\ell_I\{L\}$.

Proof: S_1 is precisely the Nielsen kernel of the surface S in the theorem. ///

(3.2) Fenchel-Nielsen coordinates on Teichmüller space

We proceed to show that decomposition into pants may be used to parametrize the Teichmüller space.

Definition: Let S be a Riemann surface of non-excluded type (g,n,m). A decomposition $\mathcal{P}$ of S into pants $\{S_1,\ldots,S_N\}$ is called a maximal (geodesic) partition if each component of $\partial_S S_i$ is a geodesic simple loop.

Implicit in the definition of maximal partition is sewing data. Specifically we know from $\mathcal{P}$ that a component C of ∂S_i is either an ideal boundary curve or puncture or that it is identified with a specific component of ∂S_j, for some $1 \leq j \leq N$, with specified orientation. (See Figure 3)

Let S have a non-excluded type and assume $\mathcal{P} = \{S_1,\ldots,S_N\}$ is a maximal decomposition of S. We label the border components

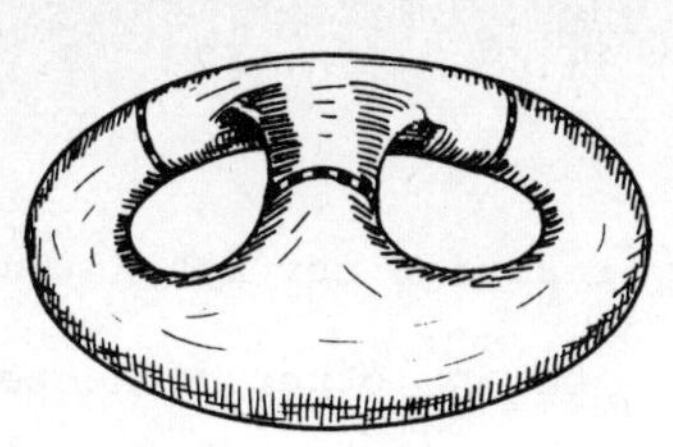 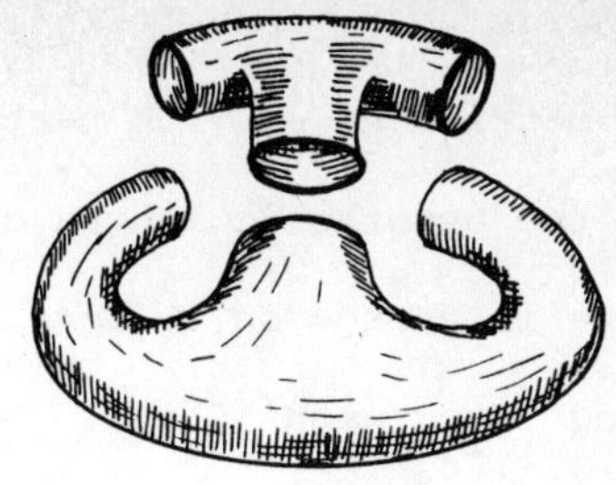

Figure 3

A maximal partition of a surface of genus 2.

of S_i by C_{ik}. For convenience set $C_{i4} = C_{i1}$. For each (j,k), in the intrinsic metric, there is a unique geodesic arc α so that:

(i) one of the endpoints of α lies on $C_{i,j}$ and the other lies on $C_{i,j+1}$,

(ii) if C_{ij} and $C_{i,j+1}$ are border curves, $\ell_I(\alpha)$ is minimal among all curves satisfying (i),

(iii) if C_{ij} (respectively $C_{i,j+1}$) is a puncture and $C_{i,j+1}$ (respectively $C_{i,j}$) is a border curve, α is orthogonal to $C_{i,j+1}$ (respectively $C_{i,j}$) .

Let ζ_{ij} denote the endpoint of α on $C_{i,j}$. ζ_{ij} is a unique <u>basepoint</u> on C_{ij}. If C_{ij} is identified with $C_{i',j'}$ choose a positive direction along C_{ij} consistent with the identification. If $\mathcal{P}$ partitions S so that C_{ij} is identified with $C_{i'j'}$, identify C_{ij} with $C_{i'j'}$ so that ζ_{ij} is identified with $\zeta_{i'j'}$. This gives a new Riemann surface S_0. Choose any marking on S_0.

Let C be any rectifiable curve on S_i with intrinsic length

$\ell_I(C)$. Notice that $\ell_I(C)$ is the same as the length of C in the intrinsic metric on S_0. In particular if S_0 has type $(g,n,0)$, $\ell_I(C)$ is the hyperbolic length of C.

There is no reason to continually restrict our attention to identifying ζ_{ij} with $\zeta_{i'j'}$. Let S' be any other connected sum of the S_i as above but not necessarily identifying ζ_{ij} with $\zeta_{i'j'}$. Assume $i < i'$ or if $i = i'$, $j < j'$. Let δ_{ij} be the oriented intrinsic distance from ζ_{ij} to $\zeta_{i'j'}$ and

$$\theta_{ij} = 2\pi\delta_{ij}/\ell_I(C_{ij}).$$

Notice that θ_{ij} is unique modulo 2π.

We may also change the intrinsic lengths of the curves C_{ij} and thereby the conformal structure of the pants S_i. If we do so in a fashion consistent with the identifications, we may again perform the gluing process. Let L be the list of the intrinsic lengths of the C_{ij} subject to the following conditions:

(i) $\ell_I(C_{ij})$ precedes $\ell_I(C_{i'j'})$ if $i < i'$ or $i = i'$ and $j < j'$,

(ii) $\ell_I(C_{i'j'})$ is omitted if $C_{i'j'}$ is identified with C_{ij} and $i < i'$ or $i = i'$ and $j < j'$,

(iii) $\ell_I(C_{ij})$ is omitted if C_{ij} is a puncture,

(iv) the lengths of dissecting curves precede the lengths of border curves.

Further let Θ be a list of the angles θ_{ij} subject to conditions (i) and (ii). We also omit θ_{ij} if C_{ij} is an ideal boundary component of S_0. L and Θ may be viewed as sets of functions on $T(S_0)$. An elementary combinatorial argument shows that, if S_0 has non-excluded type (g,n,m), a maximal partition has $2g - 2 + m + n$ elements and L has $3g - 3 + 2m + n$ elements. No angle θ_{ij} is associated to m of the curves C_{ij} whose lengths are listed in L, hence Θ has $3g - 3 + m + n$ elements. If we denote by L^{-1} the list of inverses of the numbers in L, we obtain a map

$$(L^{-1},\Theta):T(S_0) \to (R^+)^{3g-3+2m+n} \times (S^1)^{3g-3+m+n} \quad .$$

we proceed to show

<u>Lemma 1</u>: L^{-1} is a real analytic function on $T(S_0)$.

<u>Proof</u>: Let $S \in T(S_0)$ and $\alpha \subset S_0$ be a closed curve. Further assume $S = U/G$ with G normalized. If $\ell_I\{\alpha, S\}$ denotes the intrinsic length of the geodesic in the free homotopy class of α in the Poincaré metric on S, then by Lemma 2.3 $\ell_I\{\alpha, S\}$ is a real analytic function of the matrix entries of any $\gamma \in G$ covering α. The matrix entries of γ are rational functions of the matrix entries of the generators of G. The latter are almost precisely the Fricke coordinates which are real analytic on $T(S_0)$. ///

<u>Lemma 2</u>: Θ is a real analytic function on $T(S_0)$.

<u>Proof</u>: Let $S \in T(S_0)$ and $\pi:U \to S = U/G$ with G normalized. Each coordinate $\theta_i(S)$ is measured using the length of $\{\alpha\}$ and the directed distance d along $\{\alpha_i\}$ between two points x and y on α_i.

Here $\{\alpha_i\}$ is the intrinsic geodesic in the free homotopy class of α. Using the same argument as in the previous lemma, $\ell_I\{\alpha_i\}$ is a real analytic function on $T(S_0)$. d is given as an integral along a curve C in $\pi^{-1}(\{\alpha_i\})$ whose endpoints $\tilde{x}$ and $\tilde{y}$ cover x and y respectively then taking the remainder mod $2\pi\ell_I\{\alpha_i\}$. The curve is real analytically determined by $\tilde{x}$ and $\tilde{y}$. Thus the real analyticity of Θ will follow from showing that $\tilde{x}$ and $\tilde{y}$ are real analytic functions of the Fricke coordinates.

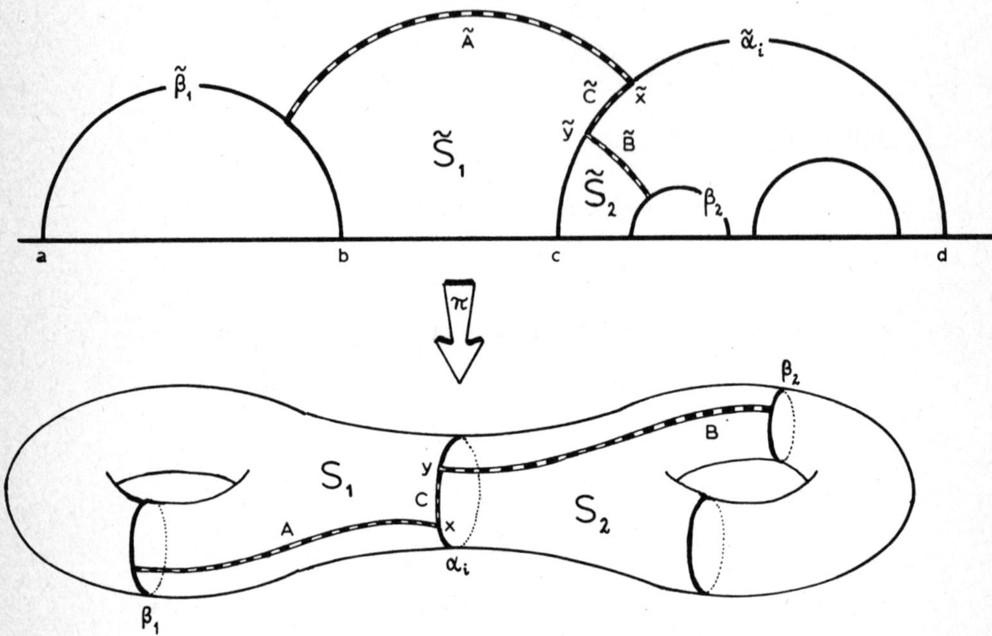

Figure 4

α_i is a border component of one or two pants in the decomposition of S into pants. Choose a fixed connected lift $\tilde{\alpha}_i$ of α_i. x (respectively y) is the endpoint of a curve A (respectively B) of shortest length connecting α_i to a geodesic β_1 (respectively β_2). We lift to U by placing tildas on everything in sight. (See Figure 4.) The endpoints of $\tilde{\beta}_j$ are the fixed points of well-defined elements of G, hence are real analytic functions on $T(S_0)$. $\tilde{A}$ is then the shortest curve from $\tilde{\beta}_1$ to $\tilde{\alpha}_i$ and $\tilde{x}$ is the endpoint on α_i. We claim x is a real analytic function of a, b, c, d. If so, by the above arguments, we are done.

Let $\eta \in \text{Aut } U$ be the unique automorphism so that $\eta(a) = \infty$, $\eta(b) = 0$ and $\eta(c) = 1$. η preserves distance and is real analytic. By symmetry of reflection in the line $\{\text{Re } z = 0\}$ we see that $\eta(A)$

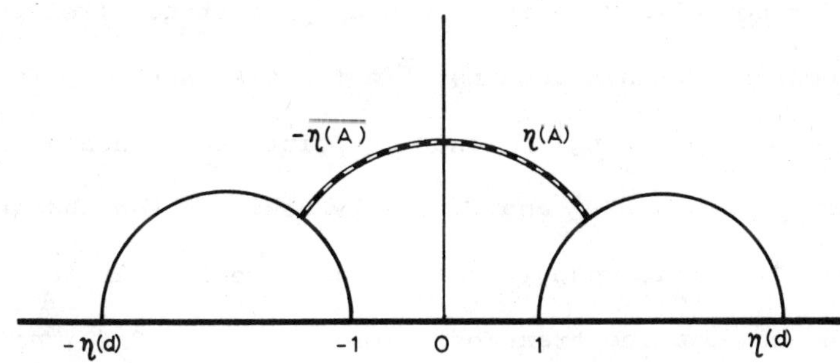

Figure 5

must be orthogonal to $\{\text{Im } z = 0\}$ and to $\eta(\alpha_i)$. The intersection of such circular arcs is clearly a real analytic function of the endpoints. Notice that we have omitted the case where the pants adjacent to α_i are tight - that argument is completely elementary.

///

<u>Theorem</u>: (L^{-1},Θ) <u>is a real analytic covering</u>.

<u>Proof</u>: The above two lemmas show that (L^{-1},Θ) is real analytic. We have already observed that it is surjective. It remains to find an evenly covered neighborhood N of $(L^{-1},\Theta)(S)$ for any $S = \chi_*(S_0) \in T(S_0)$, and on N to find a real analytic inverse to (L,Θ).

Let $\{S_1, \ldots, S_N\}$ be the decomposition of S into pants giving the Fenchel-Nielsen coordinates. Further, let S_i be the Nielsen kernel of U/G_i. From knowing $\ell_I(\alpha_{ij})$ when a_{ij} is the waist, right cuff and left cuff of S_i, we know G_i up to conjugation in Aut U.

As in Figure 4, let S_i be a component of $\pi^{-1}(S_i)$. The transformation γ covering α_i and having fixed points c and d lies both in G_1 and G_2. For surfaces $S' = \chi_*'(S_0)$ near S, assume we have renormalized so that b, c, d persist as fixed points of the corresponding transformations. It follows that G_2 is known up to a transformation fixing c and d pointwise. Such a conjugation moves y along α_i and precisely achieves the change of angle (not mod 2π). This change in angle is a real analytic function of the conjugating transformation, hence of G_2. The converse is also true. The renormalization is a real analytic diffeomorphism near S. This is precisely the gluing operation across dividing curves as realized in terms of groups, i.e., Fricke coordinates.

We must also show that gluing across non-dividing curves also gives real analytic perturbations of the cover group. Figure 6 shows the relevant picture.

We again assume a normalization with b, c, d fixed. The fundamental group of $S_1 \cup \{\alpha_j\}$ is given by Van Kampen's Theorem and is generated by $\pi_1 S_1$ and [A]. If $\eta \in G$ covers A, η is defined up to left multiplication by a transformation η_1 fixing c and d pointwise. η_1 realizes the angle of gluing real analytically.

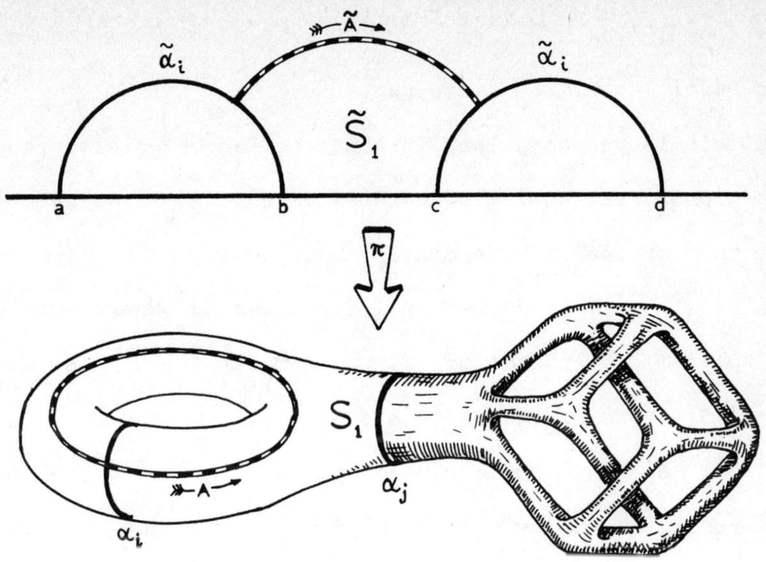

Figure 6

These gluing operations allow us to find a real analytic inverse to (L^{-1},Θ) on a neighborhood of $(L^{-1},\Theta)(S)$ defined by $|\theta(S') - \theta(S)| < \pi$. The proof is complete. ///

The above proof is a geometric van Kampen's theorem in the universal cover. It allows us to combine Fuchsian groups in a geometrically relevant form. The proof hides application of the Maskit combination theorems (c.f., [7]).

<u>Corollary</u>: The map

$$(L^{-1},\Theta):T(S_0) \longrightarrow (\mathbb{R}^+)^{3g-3+2m+n} \times (\mathbb{R})^{3g-3+m+n}$$

which covers (L^{-1},Θ) is a real analytic diffeomorphism.

<u>Definition</u>: $B(S_0) = (L^{-1},\Theta)(T(S_0))$ is the (<u>incomplete</u>) <u>Bers moduli space</u>. For $S \in T(S_0)$, the parameters $(L^{-1},\Theta)(S)$ are called the <u>Fenchel-Nielsen coordinates</u> of S.

The coordinates (L^{-1},Θ) and (L^{-1},Θ) depend on the initial choice of partition of S_0 into pants.

Implicit in the corollary is that the Fenchel-Nielsen coordinates are global real analytic coordinates on $T(S_0)$ which are unique up to a choice of basepoint.

Since (L^{-1},Θ) is a real analytic covering, there exists a group Γ of diffeomorphisms of $T(S_0)$, $\Gamma \simeq (\mathbb{Z})^{3g-3+m+n}$ so that $T(S_0)/\Gamma \simeq B(S_0)$. We now proceed to show that $\Gamma \subset \text{Mod } S_0$. For any $S \in T(S_0)$ and any simple closed geodesic α on S we form a diffeomorphism of S by splitting S along α and regluing after a rotation of 2π. (See Figure 7).

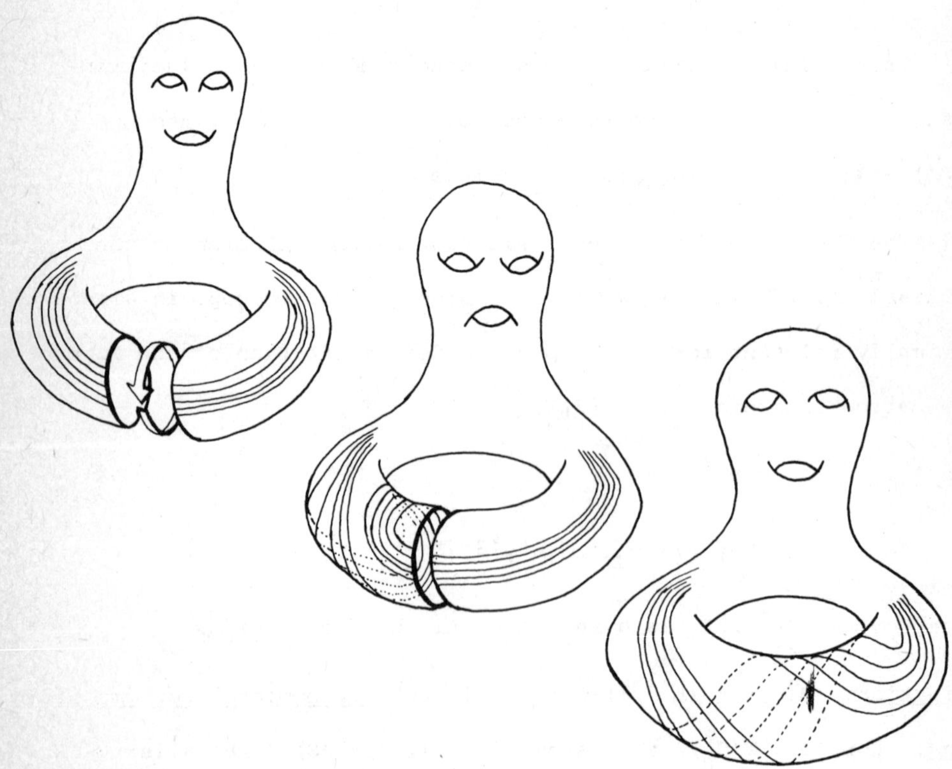

Figure 7

The isotopy class of this diffeomorphism is called the (Dehn or Lickorish) twist τ_α. In the future we will not distinguish between the diffeomorphism inducing τ_α and its class in $M(g,n,m)$ or $Mod(g,n,m)$.

If S is partitioned into pants $\{S_1,\ldots,S_N\}$ and $\alpha \subseteq \partial S_i$ is not retractable into the ideal boundary of S then the gluing $\pm 2\pi$ (the sign depending on the orientation of α) is achieved by τ_α. Thus Γ is the group $\mathcal{T}$ of twists about such curves α. The isotopy class of τ_α in $M(g,n,m)$ or $Mod(g,n,m)$ is clearly non-trivial. We then have

Theorem 2: Let S_0 be a Riemann surface of non-excluded type and $\{S_1,\ldots,S_N\}$ be a partition of S_0 into pants. Let $\alpha_1,\ldots,\alpha_k$ be a list of the boundary components of the S_i which are not retractable into the ideal boundary of S. If $B(S_0)$ is the Bers moduli space associated to this partition then $B(S_0) = T(S_0)/\mathcal{T}$ where $\mathcal{T}$ is the free abelian group generated by $\tau_{\alpha_1},\ldots,\tau_{\alpha_k}$.

(3.3) Three basic lemmas on the geometry of Riemann surfaces

Pants decompositions of Riemann surfaces of non-excluded type have been known for over 80 years. It is remarkable that many of the basic geometric properties of the surfaces and their decompositions have been discovered within the past ten years. The three results that we prove in this section are useful in compactifying the Riemann space of hyperbolic surfaces of finite conformal type. The first shows that "short" simple closed curves do not intersect.

Lemma 1: Let S be a Riemann surface of non-excluded type and α_2 a geodesic simple loop.

Let α_1 be a geodesic loop which crosses α_2. Then

$$e^{\ell(\alpha_2)} \geq \frac{e^{2\ell(\alpha_1)} + 1}{(e^{\ell(\alpha_1)} - 1)^2}$$

The bound remains valid if Poincaré length is replaced by intrinsic length. Further

(i) $\ell(\alpha_2) \to \infty$ as $\ell(\alpha_1) \to 0$

(ii) $\ell_I(\alpha_2) \to \infty$ as $\ell_I(\alpha_I) \to 0$.

Proof: Let $\pi : U \to U/G = S$ be the universal cover map. By conjugation we may assume α_2 is covered by $\gamma_2 : z \mapsto \lambda^2 z$ where $\ell(\alpha_2) = \log \lambda^2 > 0$ and $\gamma_1 : z \mapsto \dfrac{(B-k)z + B(k-1)}{(1-k)z + (kB-1)}$ covers α_1. γ_1 fixes 1 and B and $\ell(\alpha_1) = \log k > 0$. (Notice that γ_1 is not normalized.) We assume that the geodesic from 1 to B crosses $L = \{\mathrm{Re}\, z = 0\}$ hence $B > 0$. $\gamma_1(\infty) = (B-k)/(1-k) > 0$. Since L intersects none of its translates under G_1, $1 > \gamma_1(0) > 0$. Similarly $\gamma_2\gamma_1(0) > \gamma_1(\infty) > 0$ or

$$\frac{\lambda^2 B(k-1)}{kB-1} > \frac{B-k}{1-k}$$

or

$$\begin{aligned} -\lambda^2 B(k-1)^2 &> (B-k)(kB-1) \\ &> -k^2 B - B \quad . \end{aligned}$$

Thus

$$\lambda^2 \geq \frac{(k^2+1)}{(k-1)^2} \quad .$$

Conclusion (i) is then clear. To obtain the same results about intrinsic length, first double S and repeat the above argument with $\pi':U \to U/G' = S^d$. ///

We have the immediate

<u>Corollary</u>: There is a universal constant M so that if α_1 and α_2 are geodesic simple loops on a Riemann surface S of non-excluded type and $\ell(\alpha_1)$, $\ell(\alpha_2) < M$ or $\ell_I(\alpha_1)$, $\ell_I(\alpha_2) < M$, then $\alpha_1 \cap \alpha_2 = \emptyset$.

A simple closed geodesic loop α on a surface S has a tubular neighborhood N. N is 1-sided if α is a border curve and 2-sided otherwise. Up to homeomorphism, with $\alpha = \{|z| = 1\}$, in the first case $N = \{1 \leq |z| < 2\}$ and in the second case, $N = \{\frac{1}{2} < |z| < 2\}$. A <u>collar</u> C at α is a component of $N\backslash\alpha$. If α is a geodesic in the intrinsic metric a <u>collar of width</u> $W = W(C)$ at α is a collar at α such that the intrinsic distance between $z \in \alpha$ and the other component of ∂C is the constant W. We may also speak of the intrinsic area $A_I(C)$ of the collar C. By compactness, each simple closed geodesic has a collar of positive width and area. Each collar about α has two boundary components - one is α, the other we denote α^*. Our next result, the "collar Lemma" asserts that lower bounds for W and $A_I(C)$ and an upper bound for $\ell_I(\alpha^*)$ are determined solely by $\ell = \ell_I(\alpha)$. Sharp constants have been given by Matelski [8].

<u>Lemma 2</u>: Let $S = U/G$ be a Riemann surface of non-excluded type and α a border curve of S. If ℓ denotes the intrinsic length of α, then there exists constants $A = A(\ell)$, $W = W(\ell)$ and $L = L(\ell)$ so that α has a collar C of width W, area A and $\ell(\alpha^*) = L$.

Further $\lim_{\ell \to 0} W(\ell) = 0$ and $\lim_{\ell \to 0} L(\ell)$ is bounded above by $A_I(S)$.

Proof: Assume α is covered in G by the Möbius transformation $\gamma: z \mapsto \lambda^2 z$ with $\lambda = \exp(\ell_I(\alpha)/2)$. By symmetry we may assume that a collar C about α lifts to a sector $C = \{\psi_0 < |\arg z| < \pi/2\}$. The only identifications in C are given by powers of γ. In terms of ψ_0

[3.1] $$W(C) = \int_i^{\lambda i} \frac{r d\theta}{r \cdot \cos \theta} = \log(\sec \psi_0 + \tan \psi_0)$$

and

[3.2] $$A_I(C) = (\ell_I(\alpha)) \tan \psi_0 .$$

Further

[3.3] $$\ell_I(\alpha^*) = \ell_I(\alpha) \sec \psi_0 .$$

We seek to maximize both the area and width of C, hence to maximize ψ_0. ψ_0 is clearly maximized when there is a Möbius transformation $\eta \in G$ so that $\eta(I_1) \cap I_1 \neq \emptyset$ where $I_1 = \{z \mid |\arg z| = \psi_0\}$. The endpoints of $\eta(I_1)$ are fixed points of the conjugate $\gamma_1 = \eta\gamma\eta^{-1}$ of γ. (c.f. Figure 8).

By an homothety we may assume $\eta(\infty) = 1$ and $\eta(0) = B > 1$. Assuming $I = \{\mathrm{Re}\, z = 0\}$, $\eta(I)$ is the geodesic from 1 to B. Draw the tangent to $\eta(I)$ passing through 0 as in 9. The hyperbolic distance from 1 to $\eta(I)$ is precisely 2W. Let E be the center of the circle $\eta(I) \cup \eta(I)$. Further denote by $\| \ \|$ the Euclidean length. We then have

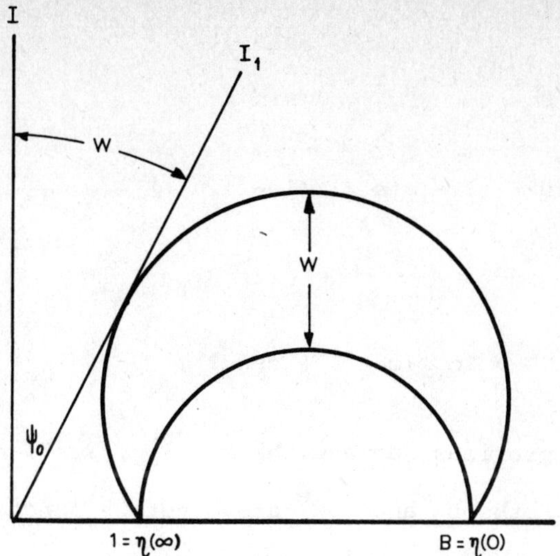

Figure 8

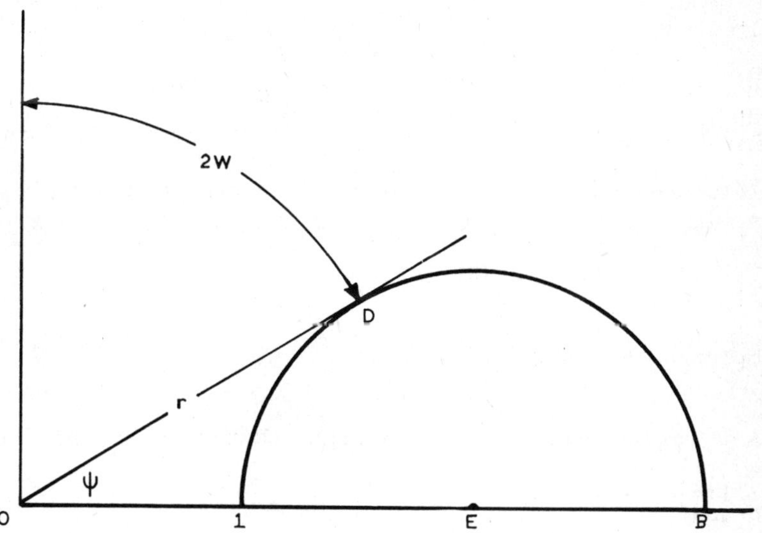

Figure 9

$$\|OD\| = r = R \cos \psi$$

$$E - 1 = \|ED\| = E \sin \psi$$

$$B = (1 + \sin \psi)/(1 - \sin \psi)$$

and

$$2W = \log (\cot \psi + \csc \psi).$$

As in the previous Lemma with $k = \lambda^2$, $\gamma_1(\infty) = (B-\lambda^2)/(1-\lambda^2) > 0$, hence $\lambda^2 > B$. Both B and $2W$ are monotone functions of ψ. Thus

$e^{\ell_I(\alpha)} = \lambda^2 > B = f(W)$. For fixed λ^2, W has a minimum which proves the existence of the constants $W(\ell)$ and $A_I(\ell)$. Clearly as $\ell_I(\alpha) \to 0$, B must tend to 1. It also follows that $\psi \to 0$ and $W \to \infty$.

For ψ_0 near $\pi/2$, $\sec \psi_0 \simeq \tan \psi_0$, hence $\ell_I(\alpha^*)$ is asymptotic to $A_I(\ell)$ near $\ell = 0$, by [3.2] and [3.3]. Since S is of finite non-excluded type and C is a subset of S,

$$A_I(C) \leq A_I(S) < \infty.$$

This completes the proof of the Lemma. ///

It is known that (c.f. Keen [5]) that a collar may be defined so that $\lim_{\ell \to 0} A(\ell) = 1$.

<u>Lemma 3</u>: There exists a constant M depending only on the type of S and the intrinsic lengths of the border curves of S with the following property:

there is a pants decomposition $\mathcal{P} = \{S_1,\ldots,S_k\}$ of S so that every border curve $\alpha \subset \partial S_i$ satisfies $\ell_I(\alpha) < M$.

Proof: If S is a pants, we are done. So assume not. Let $\alpha_1,\ldots,\alpha_m$ be the geodesic border curves of S in the intrinsic metric. It suffices to find one more curve along which to dissect S, and whose length is bounded by a number depending only on the type of S and $\{\ell_I(\alpha_i) \mid i = 1,\ldots,m\}$. Let W_i be the width of the collar at α_i given by the previous Lemma.

Let $\epsilon = \min_i W_i/2$. If there is a non-trivial simple loop α with $\ell_I(\alpha) < \epsilon$ and not retractable into ∂S, we are done. If not, every simple loop α, not retractable into ∂S and satisfying $\ell_I(\alpha) < \epsilon$, is null-homotopic. Choose a simple loop β lying in $S \setminus \{\text{collars at } \alpha_i\}$ which is not retractable into ∂S. Set $\ell = \ell_I(\beta)$. Along β, at intervals of distance 3ϵ, draw discs D_i of radius ϵ. By choice of ϵ, these discs are contained in S. The number of discs is approximately $\ell/3\epsilon$ and, for ϵ small, the area of each disc is $\pi\epsilon^2$. Thus the sum of the areas of the discs is approximately $\ell\epsilon$. By the Gauss-Bonnet Theorem, the intrinsic area $A_I(S) < \infty$ and is determined by the type of S. Thus either

(i) $\ell < A_I(S)/\epsilon$

or

(ii) the discs are not disjoint.

(i) leads immediately to the desired result. (ii) implies that in $\beta \cup (\cup D_i)$ there are closed loops β' and β'' and β is freely homotopic to some product of those loops.

Case 1: If $g > 0$, choose β to be a simple loop around a handle. In this case, β is not freely homotopic to a product of the border curves, so inductively we find a simple loop $\beta' \subset \beta \cup (\cup D_i)$ with $\ell_I(\beta') < A_I(S)/\epsilon$ and not retractable into ∂S. (To formalize this argument, notice that the homology class of β is non-trivial, but any product of curves homotopic to border curves is trivial in homology.)

Case 2: If $g = 0$, $\pi_1 S$ is freely generated by the homotopy classes of the border curves and small loops around the punctures. Deleting the collars around the border curves and also deleting horocyclic neighborhoods of the punctures gives a subsurface S' of S. Let d' be the intrinsic distance function for S restricted to S'. Further let d be the minimal d' distance between the border curves of S'. d is attained along some intrinsic geodesic β_1. Along β_1

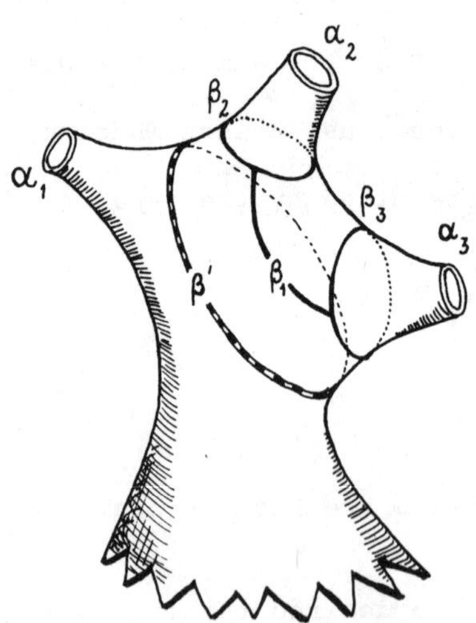

Figure 10

at distance 3ϵ draw discs D_i of radius ϵ. As before, ϵ depends only on the intrinsic lengths of the border curves. The D_i must be disjoint, hence $d\epsilon < A_I(S)$. Thus d is bounded by a constant depending only on the initial data. β_1 goes from one border curve β_2 of S' to another, say β_3. By the previous Lemma, $\ell_I(\beta_i) < C$ for $i = 2,3$, with C depending only on $\ell_I(\alpha_i)$, $i = 1,\ldots,m$. (c.f. Figure 10). $\beta' = \beta_1\beta_2\beta_1^{-1}\beta_3$ is a closed curve on S homotopic to a simple geodesic loop β. Since $\ell_I(\beta') > \ell_I(\beta)$ and $\ell_I(\beta')$ depends only on the given data, the proof is complete. ///

(3.4) Riemann surfaces with nodes and augmented moduli spaces

In choosing the Fenchel-Nielsen coordinates, we specified that the reciprocals of certain lengths are more natural coordinates than the lengths themselves. Assume α_i is some closed curve bounding a pants S_i in a pants decomposition of S. and further assume that α is not retractable into an ideal boundary curve of S. If S_i is the Nielsen kernel of U/G_i, then a change of the conformal structure of S, so that $\ell_I(\alpha_i) \to 0$, forces the ideal boundary component of S_i defined by α_i, to become a puncture. Using the results of the previous section, we may force a change in the cover group G_i so that the transformations γ_i covering α_i become parabolic. This type of deformation of structure leads us naturally to a generalization of the concept of a Riemann surface.

Definition: A Riemann surface with nodes is a connected 1-dimensional complex analytic space S with the property that each $\zeta \in S$ has

a neighborhood filterbase $\mathfrak{F} = \{N\}$ with N holomorphically either

(i) a disc, or

(ii) $\{(z,w) \in \mathbb{C}^2 \mid zw = 0 \text{ and } |(z,w)| < 1\}$.

In the second case we say that ζ is a <u>node</u> and N is a <u>conical neighborhood</u> of ζ. $\mathfrak{n}(S)$ is the set of nodes on S. A component of $S \setminus \mathfrak{n}(S)$ is called a <u>part</u> of S. S is of <u>non-excluded type</u> if each part of S is of non-excluded type. The Poincaré metric on S is precisely the Poincaré metric on the parts of S. We may also speak of marked Riemann surfaces S with nodes, by giving a marking to $S \setminus \mathfrak{n}(S)$ and specifying the identifications which occur at the nodes.

We note that a Riemann surface S with nodes has Fenchel-Nielsen coordinates given by the Fenchel-Nielsen coordinates of the parts of S. Also a Riemann surface with nodes is the structure resulting from deforming a surface S_0 so that a finite set of disjoint simple loops, not retractable into the ideal boundary of S_0, are <u>pinched</u> - i.e., identified to points. (See Figure 11).

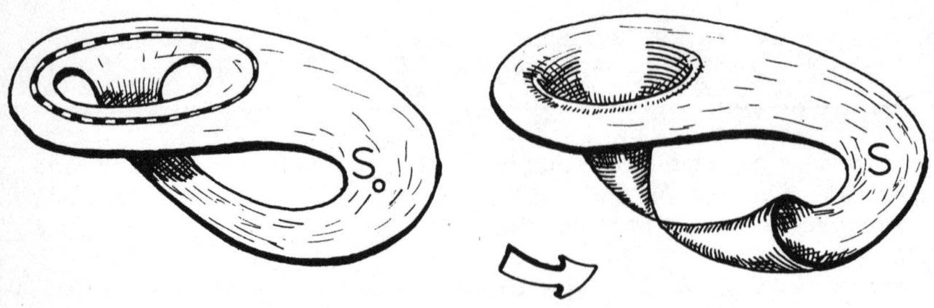

Figure 11

Before examining this degeneration phenomenon for more general surfaces, it is instructive to consider the classical case, namely type (1,1,0). A pants decomposition $\{S_1\}$ of S_0 is given by removing a single nontrivial simple loop α. The Fenchel-Nielsen coordinates for type (1,1,0) are given by the upper half-plane $U = \{z = \theta + i\ell_I^{-1}(\alpha)\}$. τ_α acts on U by translation and $B(S_0) = U/\mathfrak{T}$ is topologically a punctured disc. As $\ell_I(\alpha) \to 0$, $\operatorname{Im} z \to 0$ and, in $B(S_0)$, the deformations move into the puncture. The resulting surface S' with nodes may be used to fill in the puncture. The horocycle $\{\operatorname{Im} z > r > 0\}$ is invariant under $\mathfrak{T}$ and projects to a deleted neighborhood of S'. S' may also be viewed as the point at ∞ in U, with the horocycles as a neighborhood base at ∞.

For a marked surface S of more general non-excluded type, let $\{S_1,\ldots,S_k\}$ be a pants decomposition of S and $\{\alpha_1,\ldots,\alpha_\ell\}$ be a list of the boundary components of the S_i which are not retractable into the ideal boundary of S. We form a surface S^* with nodes by collapsing some of the curves $\{\alpha_{i_1},\ldots,\alpha_{i_p}\}$ - in the set of curves $\{\alpha_1,\ldots,\alpha_\ell\}$ - to points. Let $\{\theta_{i_1},\ldots,\theta_{i_p}\}$ be the gluing angles on S corresponding to the curves α_{i_k}, $k = 1,\ldots,p$. A <u>horocyclic neighborhood</u> $N(S^*)$ <u>of</u> (or <u>horocycle at</u>) S^* is the set of marked Riemann surfaces with or without nodes such that

(i) $\ell_I[\alpha_i, S'] = 0$ if α_i surrounds a puncture

(ii) $|\,\ell_I[\alpha_i, S'] - \ell_I[\alpha_i, S^*]\,| < \epsilon_i$ otherwise

(iii) $\theta_i(S^*)$ is arbitrary if $i \in i_1,\ldots,i_p$

(iv) $|\,\theta_i(S^*) - \theta_i(S')\,| < \delta_i$ otherwise.

Here $\ell_I(\alpha_i, S')$ and $\theta_i(S')$ denote measurement of intrinsic length and gluing angle on the surface S'. S^* is then attached to $T(S_0)$

(respectively, to $B(S_0)$ or to $R(S_0)$) using the above horocyclic neighborhoods (respectively, or their projections) as a neighborhood filterbase at S^*. The process of adding these surfaces with nodes - corresponding to all pants decompositions of S_0 - to T, B and R is called <u>augmentation</u>. The resulting spaces are the <u>augmented Teichmüller space</u>, $\hat{T}(S_0)$, the <u>augmented Bers space</u>, $\hat{B}(S_0)$, and the <u>augmented Riemann space</u>, $\hat{R}(S_0)$.

If S has type $(g,0,0)$ with $g > 1$, $R(S)$ may be identified with the moduli space M_g of stable algebraic curves of genus g (c.f. Bers [1]). It is a deep result in algebraic geometry, due in various stages to Mumford, Mayer, Knudsen and Gieseker, that M_g is a projective algebraic variety (c.f. Mumford [9]). Here we will only show that $\hat{R}(S)$ is a compact Hausdorff space. More generally, we show

<u>Theorem</u>: <u>If</u> S_0 <u>has non-excluded type</u> $(g,n,0)$, <u>then</u> $\hat{R}(S_0)$ <u>is a compact Hausdorff space</u>.

Before proceeding to the proof, we note that to the Bers space $B(S_0)$, given via some set of Fenchel-Nielsen coordinates, we may adjoin only those surfaces with nodes which realize the pinching of the boundary curves of the fixed pants decomposition of S_0. The resulting space $\tilde{B}(S_0) = B(S,\mathcal{P})$ is, by Theorem 3.2, a product of discs with a Euclidean space, hence is Hausdorff.

<u>Definition</u>: Let S_0 be as above and $\mathcal{P}$ and $\mathcal{P}'$ be pants decompositions of S_0. $\mathcal{P} = \{S_1,\ldots,S_k\}$ and $\mathcal{P}' = \{S_1',\ldots,S_k'\}$ are called <u>congruent</u> if there exists a homeomorphism or equivalently a

diffeomorphism $f\colon S_0 \to S_0$ so that $f(S_i) = S_i'$ for all $i \in \{1,\dots,k\}$.

A simple combinatorial argument shows

<u>Lemma</u>: If S_0 has non-excluded type then there are only finitely many non-congruent decompositions of S_0 into pants.

<u>Proof of the Theorem</u>: To each $S \in \hat{R}(S_0)$, using Lemma 3.3.3, we may associate a pants decomposition $\mathcal{P} = \{S_1,\dots,S_k\}$ so that $\ell(\alpha) < L_0$ for each border curve α of one of the S_i. We define a neighborhood filterbase $\mathcal{F}_0 = \{N_\mu(S)\}$ at S. $N_\mu(S)$ is the set of $S' \in \hat{R}(S_0)$ which admit a pants decomposition $\mathcal{P}' = \{S_1',\dots,S_k'\}$ congruent to $\mathcal{P}$ with S_i corresponding to S_i' and such that, for all border curves α_{ij}' of S_i' corresponding to α_{ij} on S_i

(N1) $\qquad \ell(\alpha_{ij}') = 0$ if α_{ij} represents a puncture not a node

(N2) $\qquad |\ell(\alpha_{ij}') - \ell(\alpha_{ij})| < 1/\mu$,

and the gluing angles θ_{ij} and θ_{ij}' satisfy

(N3) $\qquad |\theta_{ij}' - \theta_{ij}| < 1/\mu$ if $\ell(\alpha_{ij}) \neq 0$

(N4) $\qquad \theta_{ij}'$ is arbitrary otherwise.

It is obvious that $\mathcal{F}_0$ is a base for the neighborhood filter at S given earlier in this section. We have thus shown that $\hat{R}(S_0)$ is first countable.

Let $S^\nu \in \hat{R}(S_0)$ for $\nu \in \mathbb{Z}$. Choose pants decompositions $\mathcal{P}^\nu$ of S^ν so that each pants has border components $\{\alpha_{ij}^\nu\}$ with $\ell(\alpha_{ij}^\nu) < L_0$. By passing to a subsequence and using the Lemma, we may assume that the $\mathcal{P}^\nu$ are congruent to some fixed $\mathcal{P}_0$. Again, by passing to a subsequence, we may assume that $\ell(\alpha_{ij}^{\nu}) \to L_{ij}$ as $\nu \to \infty$. Modulo twist maps, the gluing angle θ_{ij}^{ν} satisfies $\theta_{ij}^{\nu} \in [0,2\pi]$. We choose a convergent subsequence with limit θ_{ij} of each θ_{ij}^{ν} (mod 2π) as $\nu \to \infty$. Thus S^{ν} has a subsequence convergent to the surface $S \in \hat{R}(S_0)$ whose Fenchel-Nielsen coordinates with respect to $\mathcal{P}_0$ are (L_{ij}, θ_{ij}). Therefore $\hat{R}(S_0)$ is compact.

Let S' and S'' be distinct points in $\hat{R}(S_0)$. We show that for μ sufficiently large $N_\mu(S') \cap N_\mu(S'') = \emptyset$. First note that $\bigcap_{\mu>0} N_\mu(S) = \{S\}$. If $S_\mu \in N_\mu(S') \cap N_\mu(S'')$, then S_{μ_j} accumulates at both S' and S'' for any sequence $\mu_j \to \infty$. But the Fenchel-Nielsen coordinates of any accumulation point are well defined. Thus S' and S'' are conformally equivalent to $\lim S_{\mu_j}$, hence to each other. This contradiction of our assumption shows that $\hat{R}(S_0)$ is Hausdorff. ///

There is a particularly elegant construction of $\hat{R}(S_0)$ which is due to W. Harvey [4].

References

1. L. Bers, On spaces of Riemann surfaces with nodes, Bull. Amer. Math. Soc. 80(1974), 1219-1222.

2. L. Bers, Nielsen extensions of Riemann surfaces, Ann. Acad. Sci. Fenn. 2(1976), 17-22.

3. H. Cartan, Elementary Theory of Analytic Functions of One and Several Complex Variables, Addison-Wesley, Reading, Mass., 1963.

4. W. Harvey, Spaces of discrete groups, Discrete Groups and Automorphic Functions, Academic Press, London, 1977, 295-348.

5. L. Keen, Collars on Riemann surfaces, Discontinuous Groups and Riemann Surfaces, Ann. of Math. Studies 79(1974), 263-268.

6. A. Macbeath, Discontinuous Groups and Birational Transformations, Proceedings of the Summer School in Mathematics, Queens College, Dundee, 1961.

7. B. Maskit, On Klein's combination Theorem, III, Advances in the Theory of Riemann Surfaces, Ann. of Math. Studies 66(1971), 297-316.

8. J. P. Matelski, A compactness theorem for Fuchsian groups of the second kind, Duke Math. J. 4(1976), 829-840.

9. D. Mumford, Stability of projective varieties, IHES, preprint.

10. C. L. Siegel, Topics in Complex Function Theory, Vol. II, Wiley, New York, 1971.

CHAPTER III

THE CLASSIFICATION OF THE DIFFEOMORPHISMS OF SURFACES OF FINITE TYPE AND GROUP ACTIONS ON TEICHMÜLLER SPACE

Thurston's study of measured foliations of surfaces and his compactification of Teichmüller space lead naturally to a classification of the diffeomorphisms of topologically finite orientable surfaces. In [1], Bers considered an extremal problem in Teichmüller theory and showed that the solution - or non-existence of a solution - leads to the same classification. Here we will present Bers' method. Still another approach to obtaining the classification is due to W. J. Harvey [4]. He applies the techniques of Tits complexes and graphs of groups to the Teichmüller modular group.

Section 4 is devoted to the study of a continuous group action on the Teichmüller space. It is analogous to the group of automorphisms of the unit disc Δ which is generated by the horocyclic flows. The group action is shown to be transitive on Teichmüller space.

The concluding remarks to these notes form §5. Several open problems are stated there.

§1 Bers' extremal problem and the computations for punctured tori

Let S be a differentiable oriented surface of finite nonexcluded type (g,n,m) and $f \in \mathrm{Diff}_+ S$. S may be given the structure of a Riemann surface S_σ by pullback from a diffeomorphism $\sigma: S \to S'$, where S' is a Riemann surface. In the homotopy class (f) of f we choose the extremal map f_σ so that

$$K[\sigma\circ f_\sigma\circ\sigma^{-1}] = \inf_{f_1 \in (f)} K[\sigma\circ f_1\circ\sigma^{-1}].$$

An immediate consequence of the Teichmüller theorems for surfaces of finite non-excluded type is that f_σ exists, is unique and $\sigma\circ f_\sigma\circ\sigma^{-1}$ is either conformal or is a Teichmüller mapping. Bers considered the problem of minimizing $K[\sigma\circ f_\sigma\ \sigma^{-1}]$ over all possible conformal structures σ. We set

$$\overline{K}(f) = \inf_\sigma K[\sigma\circ f_\sigma\circ\sigma^{-1}]$$

and consider the following two questions:

(B1) Is $\overline{K}(f) > 0$?

(B2) Is there some conformal structure σ on S so that

$$\overline{K}(f) = K[\sigma\circ f_\sigma\circ\sigma^{-1}]?$$

A conformal structure at which $\overline{K}(f)$ is attained is called f-<u>minimal</u>. Since any extremal function f_σ is a Teichmüller mapping or is conformal, the problem is posed as a question about diffeomorphism classes acting on Teichmüller space. In fact, (B2) may be rephrased as:

(B2') If $f \in \mathrm{Diff}_+ S$ induces $g_f \in \mathrm{Mod}\ S_1$ does there exist $S' \in T(S)$ so that

$$d_T(S', g_f(S')) = \inf_{S'' \in T(S)} d_T(S'', g_f(S''))?$$

The restatement is an immediate consequence of the definition of Teichmüller distance (c.f. Chapter I, §5). To classify elements g of Mod S, assume $f \in \mathrm{Diff}_+ S$ induces g. A classification is given by the following table.

	(B1) yes	(B1) no
(B2) yes	hyperbolic	elliptic
(B2) no	pseudo-hyperbolic	parabolic

This classification is a generalization of the classification of the Teichmüller modular group for the once-punctured torus. Since this case is classical and easily computed, we will sketch the argument. Let $L = L_{1,\tau}$ be a lattice in $\mathbb{C}^1$. We identify L with the points $n + m\tau$ with $n, m \in \mathbb{Z}$. We may assume $\operatorname{Im} \tau > 0$. $(\mathbb{C} \setminus L)/L$ is a punctured torus and, by the uniformization theorem, all once-punctured tori may be so obtained by considering all $\tau \in U$. U is therefore a model for $T_{1,1,0}$. Two points $\tau_1, \tau_2 \in U$ determine conformally equivalent tori if and only if $L_{1,\tau_1} = L_{1,\tau_2}$. Equivalently, $\tau_2 = (a\tau_1+b)/(c\tau_1+d)$ with $ad - bc = 1$ and $a,b,c,d \in \mathbb{Z}$. This group is the classical elliptic modular group and is the Teichmüller modular group $\operatorname{Mod}(1,1,0)$. If $g \in \operatorname{Mod}(1,1,0)$, g is either parabolic, elliptic or hyperbolic qua Möbius transformation. The classification (in the hyperbolic metric) follows immediately. The translation $z \mapsto z + 1$ is the twist map indicated in Figure 1.

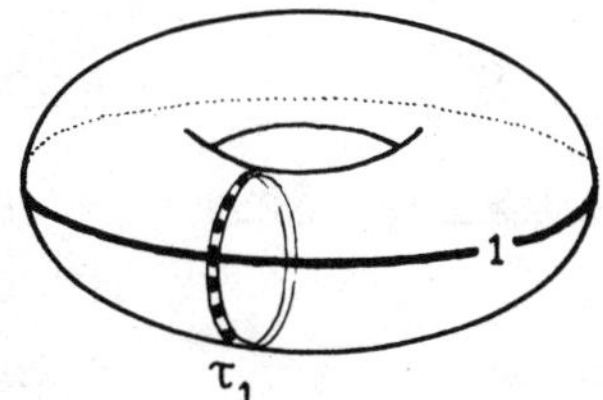

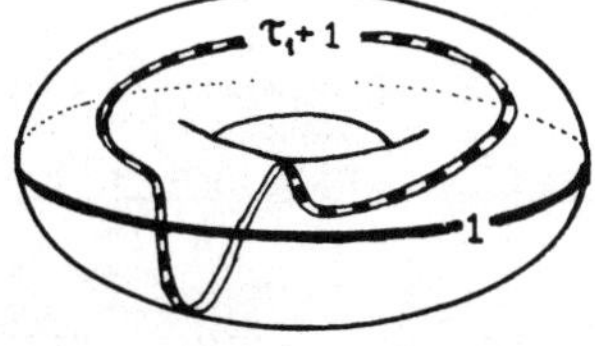

Figure 1

Mod(1,1,0) also contains rotations γ_2 and γ_3 of periods 2 and 3 respectively. These fix punctured tori with the symmetries of the period parallelogram indicated in Figure 2. Pseudo-hyperbolic transformations occur only in higher dimensional Teichmüller spaces.

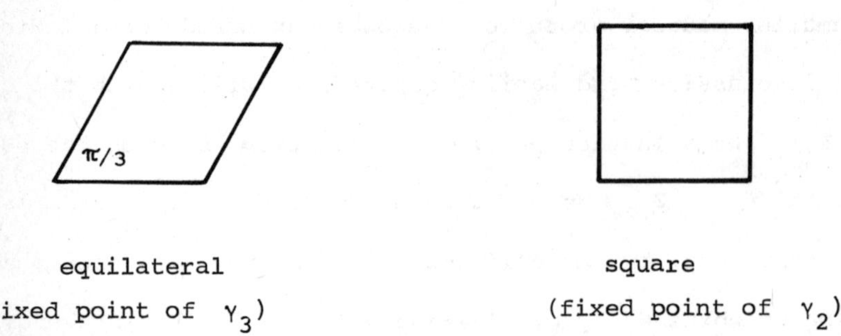

Figure 2

It is known (c.f. Kravetz [3] or Royden [6]) that d_T, the Teichmüller distance, agrees with the hyperbolic distance in T(1,1,0). In higher dimensional Teichmüller spaces, any two points lie in a unique plane which, in the Teichmüller metric, is isometric to the hyperbolic plane. This is used to endow Teichmüller space with a Finsler geometry (c.f. the above references).

We now proceed to relate the different types of modular transformations to the geometry of the diffeomorphisms that induce them.

§2 Periodic diffeomorphisms and elliptic transformations

An orientation-preserving diffeomorphism $f: S \to S$ is <u>periodic</u> if $f^N = id$ for some $N > 0$.

<u>Theorem 1</u>: f <u>is isotopic to a periodic map on</u> S <u>if and only if there is a conformal structure</u> σ <u>on</u> S <u>and a self-map</u> f' <u>isotopic to</u> f <u>so that</u> $\sigma \circ f' \circ \sigma^{-1}$ <u>is conformal</u>.

<u>Proof</u>: If $f'' = \sigma \circ f' \circ \sigma^{-1}$ is conformal on S' then f'' lies the group Aut S'. But Aut S' is finite by Theorem (2.23) of Chapter II. Thus $(f'')^N = \mathrm{id}$ for some $N \geq 0$ and $(f')^N = \mathrm{id}$.

Conversely, if f' is periodic on order N, let ds^2 be any Riemannian metric on S. We form the pullback metrics $(f')^i{}_*(ds^2)$ of ds^2 by $(f')^i$.

$$ds_0^2 = \sum_0^{N-1} (f')^i{}_*(ds^2)$$

is a Riemannian metric invariant under f'. ds_0^2 gives a conformal structure to S in which f' is conformal. ///

Notice that f acts on boundary components of S and the orbit under f^k of a puncture consists solely of punctures; similarly the orbit of a boundary curve consists of boundary curves. If ds^2 is an intrinsic metric, then ds_0^2 will have punctures precisely at points which are in the orbit of a ds^2 puncture. To see this, compute the ds_0^2 length of a curve from a point $z_0 \in S^o$ to the boundary curve.

<u>Corollary 1</u>: If $g_f \in \mathrm{Mod}(g,n,m)$ is induced by a periodic self-map f which permutes the punctures on $S \in T_{g,n,m}$ then g_f has a fixed point in $T_{g,n,m}$.

<u>Proof</u>: With respect to any marking on S_1 the conformal structure on S given by ds_0^2 admits f as a conformal self-map. The result then follows from Theorem (2.1.1) of Chapter II. ///

<u>Corollary 2</u>: Let $f \in \mathrm{Diff}_+ S$ where S has non-excluded type (g,n,m). Then f is isotopic to a periodic self-map of S which permutes the punctures on S if and only if $g_f \in \mathrm{Mod}(g,n,m)$ is elliptic.

<u>Proof</u>: If g is elliptic, then any $f \in \mathrm{Diff}_+ S$ inducing g is isotopic to a periodic self-map of the surface S_σ at which $d_T(S_\sigma, g(S_\sigma)) = 0$ by Theorem (2.1.1) of Chapter II. If f is isotopic to a periodic self-map of S which permutes the punctures on S_σ with $S_\sigma \in T_{g,n,m}$, then g_f is elliptic by Corollary 1. ///

An easy consequence of Corollaries 1 and 2 is

<u>Corollary 3</u>: A transformation $g \in \mathrm{Mod}(g,n,m)$ is elliptic if and only if it is of finite order.

<u>Theorem 2</u>: <u>Assume</u> S <u>is an orientable differentiable surface of non-excluded type, genus</u> g <u>and</u> k <u>ideal boundary components and</u> $f \in \mathrm{Diff}_+ S$. <u>Then</u> g_f <u>is elliptic and</u> $\overline{K}(f) = 0$ <u>is attained at some surface</u> $S \in T_{g,n,k-n}$ <u>if and only if</u> f <u>is isotopic to a periodic self-map of</u> S. <u>Further</u> $m = k - n$ <u>may assume the values</u> 0 <u>or</u> k <u>and possibly other values between</u> 0 <u>and</u> k <u>depending on</u> f.

It is important for us to note that diffeomorphisms f isotopic to periodic diffeomorphisms may achieve $\overline{K}(f) = 0$ at a surface S which is not of finite conformal type. By comparison we have

<u>Theorem 3</u>: <u>Let</u> $S_\sigma \in T_{g,n,m}$ <u>with</u> $m \neq 0$ <u>and</u> $f \in \mathrm{Diff}_+ S_\sigma$. <u>If</u> f <u>is not isotopic to a periodic self-map of</u> S_σ, <u>then</u> $\overline{K}(f) < K[f_\sigma]$.

<u>Proof</u>: Let S^d be the double of S_σ and $\pi\colon U \to U/G^d = S^d$ be the

universal cover of S^d. Let Δ be a connected lift of S_σ in U and G_σ be the subgroup of G^d which stabilizes Δ. $\Delta/G_\sigma = S_\sigma$ and we have noted in §1.3 of Chapter II that U/G_σ is the Nielsen extension S^N of S_σ.

Assume $f_\sigma: S_\sigma \to S_\sigma$ is the extremal map for Bers' problem. Since f_σ is a Teichmüller mapping, f_σ extends to S^d and therefore lifts to a map $F: U \to U$. $K[F] = K[f_\sigma]$. FG^dF^{-1} is conjugate in Aut U to G^d since $S_\sigma = f_\sigma(S_\sigma)$. Therefore for some $\gamma \in$ Aut U, $G_\sigma = \gamma F G_\sigma F^{-1}\gamma^{-1}$, and F covers a map $f_N: S^N \to S^N$ which is homotopic to f_σ. Notice that $f_N \mid S_\sigma = f_\sigma$ and

$$\mu_{f_N} = \mu_{f_\sigma}.$$

Since we have assumed that f_σ is extremal for Bers' problem,

$$K[F] = K[f_\sigma] \leq K[f_N] = K[F].$$

Thus f_N is also extremal for Bers' problem and is therefore a Teichmüller mapping.

We wish to show that f_σ and f_N cannot simultaneously be Teichmüller mappings. To do so, we recall that a Teichmüller mapping of a bordered surface is given by some $K > 0$ and a non-zero quadratic differential ω on S which extends to S^d holomorphically possibly with simple poles at the punctures. A routine computation then shows that $\iint_S |\omega| < \infty$, where, in local coordinates, $\omega = \varphi(z)dz^2$ and $|\omega| = 4|\varphi|\,dxdy$.

In any local coordinate z on S, $(f_\sigma)_{\bar z} / (f_\sigma)_z = \mu(z)$ determines ω up to a positive real factor since $\omega = \varphi(z)dz^2$ and $\mu = k|\varphi|/\varphi$. ω extends to a quadratic differential $\omega^d \in \mathcal{Q}_{S^d} \setminus \{0\}$.

Since f^d is a Teichmüller mapping associated to ω^d,

$$\iint_{S^d} |\omega^d| < \infty.$$

But S^N is an infinite sheeted cover of S^d and ω^N covers ω^d. It follows that

$$\iint_{S^N} |\omega_N| = \infty$$

and f_N cannot be a Teichmüller mapping. This contradicts our assumption that f^d is extremal. ///

Corollary: If S is a surface of non-excluded type and $f \in \mathrm{Diff}_+ S$ is not isotopic to a periodic self-map, then $\overline{K}(f)$ is attained only at a Riemann surface structure σ so that S_σ is of finite conformal type.

The above corollary is the key to the remainder of our study of diffeomorphisms of surfaces since it allows us to restrict our attention to Riemann surface structures S where $\hat{R}(S)$ is compact.

If $\overline{K}(f)$ is not attained, we will be forced to consider minimizing sequences, i.e. structures σ_n on S so that $K[\sigma_n \circ f_{\sigma_n} \circ \sigma_n^{-1}] \to \overline{K}(f)$. We will need to show that there is also a minimizing sequence σ_n' so that $S_{\sigma_n'}$ is of finite conformal type. The existence of this latter minimizing sequence follows directly from

Theorem 4: Let S be a differentiable surface and σ_1 a Riemann surface structure on S which is of non-excluded type. Denote by σ_∞ the Riemann surface structure on S which is the infinite Nielsen extension of S_{σ_1}. Then, for any $f \in \mathrm{Diff}_+ S$,

$$K[\sigma_\infty \circ f_{\sigma_\infty} \circ \sigma_\infty^{-1}] \leq K[\sigma_1 \circ f_{\sigma_1} \circ \sigma_1^{-1}].$$

<u>Proof</u>: As in Chapter II §1.2, $S_\infty = S_{\sigma_\infty}$ is the increasing union of S_k where S_k is the Nielsen extension of S_{k-1} and $S_1 = S_{\sigma_1}$. We showed in the proof of the previous theorem that there is a homeomorphism f_2 of S_2 so that $f_2 \mid S_1 = f_1 = \sigma_1 \circ f_\sigma \circ \sigma_1^{-1}$. Iterating this process, we obtain a map $f_\infty \colon S_\infty \to S_\infty$ and satisfying $\|\mu_{f_\infty}\| = \|\mu_{f_{\sigma_1}}\|$ where $\|\ \|$ is the essential supremum. As in Chapter I,

$$K[f_\infty] = \frac{1 + \|\mu_{f_\infty}\|}{1 - \|\mu_{f_\infty}\|} = K[f_{\sigma_1}].$$

We lift both f_∞ and f_{σ_∞} to F_∞ and F_{σ_∞} mapping U to U. First note that μ_{f_∞} may be approximated in $L^\infty(S_\infty)$ by smooth $(-1,1)$-forms ν_k on S_∞. Let $\tilde{\nu}_k$ cover ν_k in U. The solution H_k to the Beltrami equation $(H_k)_{\bar{z}} = \tilde{\nu}_k (H_k)_z$ is smooth by the Ahlfors-Bers theorem. H_k covers an admissible map $h_k \colon S_\infty \to S_\infty$ which lies in the homotopy class of $\sigma_\infty \circ f \circ \sigma_\infty^{-1}$. Since $\sigma_\infty \circ f_{\sigma_\infty} \circ \sigma_\infty^{-1}$ is the Teichmüller mapping in the homotopy class of the admissible map h_k,

$$K[\sigma_\infty \circ f_{\sigma_\infty} \circ \sigma_\infty^{-1}] \leq K[h_k] \to K[\sigma \circ f_\sigma \circ \sigma^{-1}].$$

The desired result follows immediately. ///

The above proof is rather contorted and does not give the best possible result. Using the theory of quasiconformal mappings with distributional derivatives, one may derive the above inequality directly. One may further show that strict inequality holds unless S_{σ_1} is of finite conformal type. This latter result requires an extension of the version of Teichmüller's Theorem given in these notes.

We also have the immediate

Corollary: If σ_n is a sequence of conformal structures on S so that $K[\sigma_n \circ f_{\sigma_n} \circ \sigma_n^{-1}] \to \overline{K}(f)$ then there exist conformal structures σ_n' of finite conformal type so that $K[\sigma_n' \circ f_{\sigma_n'} \circ \sigma_n'^{-1}] \to \overline{K}(f)$.

§3 Irreducible diffeomorphisms and hyperbolic transformations

Thurston introduced the concept of a reducible diffeomorphism to handle those diffeomorphisms of surfaces which restrict to (not necessarily connected) subsurfaces. In this section we shall show that diffeomorphisms isotopic to irreducible ones induce elliptic or hyperbolic transformations, and study hyperbolic transformations. We proceed to the precise definitions.

(3.1) Irreducible diffeomorphisms

Let S be a surface of non-excluded type and $\alpha_1,\dots,\alpha_k$ be a set of disjoint simple loops on S. $\{\alpha_1,\dots,\alpha_k\}$ properly decomposes S if each component of $S \setminus \bigcup \alpha_i$ is of non-excluded type. If $\{\alpha_1,\dots,\alpha_k\}$ properly decomposes S and $f \in \mathrm{Diff}_+ S$ satisfies $f(\cup\alpha_i) = \cup\alpha_i$ then we say that f is reduced by $\{\alpha_1,\dots,\alpha_k\}$. f is irreducible if it is not isotopic to a reducible diffeomorphism.

We give two examples of reducible diffeomorphisms. Let

$$S_1' = \{z \mid |z| < 2,\ |z - e^{in\pi/2}| > 1/2,\ n = 1,2,3,4\}$$

S_1' admits a period four conformal self-map $z \mapsto ze^{i\pi/2}$ which extends to a conformal self-map f_1 of $S_1 = (S_1')^d$. f_1 is clearly reducible. Another reducible diffeomorphism on any surface S of positive genus is given by the twist map about a simple loop α which is homotopically non-trivial. It is more difficult to give examples of irreducible diffeomorphisms now. We can however prove

Theorem 1: *If f is irreducible then g_f is hyperbolic or elliptic.*

Proof: If $\overline{K}(f)$ is attained at $S_\sigma \in T(S)$ we are done. We may therefore assume S_{σ_n} is a minimizing sequence for $\overline{K}(f)$. By Theorem 4 of the previous section, we may assume σ_n is a conformal structure on S which is of finite conformal type. We previously viewed S_{σ_n} as an element of $T(S_{\sigma_0})$ where S_{σ_0} is a marked Riemann surface structure on the surface S of finite conformal type. The problem however is independent of the marking hence is well-posed on $R(S_0) = \pi(T(S_0))$. By continuity of the Teichmüller distance, if $\pi(S_{\sigma_n}) \to S_\infty \in R(S_0)$, S_∞ is an f-minimal conformal structure on S, contrary to hypothesis. If $\pi(S_{\sigma_n})$ diverges in $R(S_0)$, then, by compactness of $\hat{R}(S_0)$, on a subsequence $\pi(S_{\sigma_{n_k}})$ converges to a Riemann surface S_∞ which has nodes.

Now $K[\sigma_k \circ f_{\sigma_k} \circ \sigma_k^{-1}] \to \overline{K}(f)$. So we may assume $K[h_k] < M < \infty$ where $h_k = \sigma_k \circ f_{\sigma_k} \circ \sigma_k^{-1}$. If $\alpha_1, \ldots, \alpha_i$ are simple loops on S_0 which are pinched on S_∞, then $\ell(\alpha_j, S_{\sigma_k}) \to 0$ as $k \to \infty$ for $j = 1, \ldots, i$. By Wolpert's Theorem (Chapter II, Theorem 1.3.3) the geodesic in the class of $f(\alpha_j)$ on S_{σ_k} has length $\ell_{j,k} < K[h_k] \cdot \ell(\alpha_j, S)$. Thus $\ell_{j,k} \to 0$ for all j as $k \to \infty$. By Lemma 3.3.1 of Chapter II, the α_j are disjoint and their homotopy classes must be permuted by f. A smoothed PL-map f_1 isotopic to f will then permute the α_j and

f is not irreducible. ///

(3.2) The metric geometry of Teichmüller space

We have noted in Chapter I that the Teichmüller space carries the natural geometry of a "straight space" in the sense of Busemann. We shall need to consider some of the notions of this geometry. Let (X,δ) be a metric space. A geodesic segment in X is an arc in X along which the distance function δ is uniquely additive. Precisely, if $x_1, x_2 \in X$ and $\delta(x_1,x_2) = d$, let $j: [0,d] \to X$ be continuous with $j(0) = x_1$ and $j(d) = x_2$. If $\delta(j(r),j(s)) = |r - s|$, $G = j([0,d])$ is called a geodesic segment between x_1 and x_2. G is a line segment between x_1 and x_2 if G is the unique geodesic segment between x_1 and x_2. $G_1 \supset G$ is a line through x_1 and x_2 if j extends to a mapping $j': \mathbb{R} \to X$ with $j'(\mathbb{R}) = G_1$ and $j'(a,b)$ a line segment in X for all $a < b$. (X,δ) is a straight space (in the sense of Busemann) if, for each $x_1, x_2 \in X$, with $x_1 \neq x_2$, there is a line through x_1 and x_2.

We wish to show that the Teichmüller space $T(g,n,0)$ together with the Teichmüller metric d_T is a straight space. There is a natural candidate for the line passing through two distinct points S_0 and S_1. We base the Teichmüller space at S_0. Then the Teichmüller deformation

$$i: S_0 \mapsto \left(S_0, \frac{\omega}{||\omega||}, ||\omega||\right) = S_1 \quad \text{for } \omega \in Q_{S_0} \setminus \{0\}$$

satisfies

$$d_T(S_0,S_1) = \log \frac{1 + ||\omega||}{1 - ||\omega||}.$$

Let $\tau = (1+t)/(1-t)$ for $t \in]-1,1[$ and

$$S_\tau = \begin{cases} (S_0, \omega/\|\omega\|, t) & \text{for } t \geq 0 \\ (S_0, \omega/\|\omega\|, -t) & \text{for } t \leq 0 \end{cases}$$

with $t \in]-1,1[$. Further, let

$$j' : \mathbb{R} \to T_{g,n,0}$$
$$\tau \mapsto S_\tau .$$

$j'(\mathbb{R})$ is called a <u>Teichmüller line</u> through S_0.

<u>Lemma 1</u>: j' is an isometry.

<u>Proof</u>: By the Teichmüller theorems j' preserves distance from S_0. Assume $b > a$. Let $S_a = j'(a)$ and $S_b = j'(b)$. Then

$$i_1 : S_0 \to \left(S_0, \frac{\omega}{\|\omega\|}, \frac{e^a-1}{e^a+1}\right) = S_a$$

and

$$i_2 : S_0 \to \left(S_0, \frac{\omega}{\|\omega\|}, \frac{e^b-1}{e^b+1}\right) = S_b$$

Let $\zeta = \xi + i\eta$ be the ω-coordinate on S_0. Then

$$\zeta_1 = \xi_1 + i\eta_1 = e^a\xi + i\eta$$

is the local coordinate on S_a induced by the Teichmüller deformation i_1. Similarly, $\zeta_2 = \xi_2 + i\eta_2 = e^b\xi + i\eta$ is the induced local coordinate on S_b. Set $\omega_1 = d\zeta_1^2$. The Teichmüller deformation

$$i_\zeta : S_0 \to \left(S_0, \frac{\omega_1}{\|\omega_1\|}, \frac{e^{b-a}-1}{e^{b-a}+1}\right)$$

has induced local coordinate

$$\zeta_3 = e^{b-a}\xi_1 + i\eta_1 = e^b\xi + i\eta = \zeta_2 .$$

Thus $d_T(S_a, S_b) = b - a$ and j' is an isometry. ///

We may now show

<u>Theorem</u>: $(T_{g,n,0}, d_T)$ <u>is a straight space</u>.

<u>Proof</u>: By Teichmüller's Existence Theorem, given any two distinct $S_0, S_1 \in T_{g,n,0}$, there is a Teichmüller line through S_0 and S_1. Each subarc of a Teichmüller line L is a geodesic segment by the previous Lemma. To show that $T_{g,n,0}$ is a straight space, it is only necessary to show that each geodesic segment G in L is a line segment. To do so, it suffices to show that if S_1 and S_2 are the endpoints of G and $S_3 \in T_{g,n,0}$ satisfies

$$d_T(S_1, S_2) = d_T(S_1, S_3) + d_T(S_3, S_2)$$

then $S_3 \in G$. For $\ell = 1,2$ set $\log K_\ell = d_T(S_\ell, S_3)$ and $k_\ell = (1-K_\ell)/(1+K_\ell)$. Further, let $\log K_3 = d_T(S_1, S_2)$ and $k_3 = (1-K_3)/(1+K_3)$. Since G is a geodesic segment, there is a surface $S' \in G$ so that $d_T(S_\ell, S') = \log K_\ell$ for $\ell = 1,2$. In particular, we have

$$[1] \qquad k_3 = \frac{k_1+k_2}{1+k_1k_2} .$$

For $p,q = 1,2,3$ let f_{pq} be the Teichmüller deformation from S_p to S_q. Again, as in Chapter I, $f_{qp} = f_{pq}^{-1}$ and $\log K[f_{pq}] = d_T(S_p, S_q)$. By Teichmüller's Uniqueness Theorem, equation

[1] implies $f_{12} = f_{32} \circ f_{13}$ or $f_{32} = f_{12} \circ f_{13}^{-1}$.

As in the proof of Lemma (3.5.2) of Chapter I

$$\mu_{f^{-1}}(f(z)) = -\mu_f(z) \exp\{2i \arg f_z(z)\}$$

and

$$\arg (f^{-1})_z \Big|_{f(z)} = -\arg (f)_z \Big|_z$$

for any admissible map f and a local coordinate z. Set $f = f_{32}$ and $w = f_{31}$. Except at a finite number of points on S_3, the above computations yield

[2] $$\mu_{f\circ w^{-1}}(z) = \frac{\mu_f(z) - \mu_w(z)}{1 - \mu_f(z)\overline{\mu}_w(z)} \{\exp 2i \arg w_z(z)\}$$

for some holomorphic local coordinate z on S_1. Since f_{32}, f_{31} and f_{12} are Teichmüller mappings, $|\mu_{f_{12}}| \equiv k_3$ and there exist $\omega_i = \varphi_i dz^2 \in Q_{S_3} \setminus \{0\}$ with $\|\omega_i\| = 1$ so that

$$\mu_{f_{3i}} = k_i \frac{|\varphi_i|}{\varphi_i} \quad \text{for } i = 1,2.$$

Then Equation [2] shows that

$$k_3 = |\mu_{f_{12}}| = \left| \frac{k_1 \dfrac{|\varphi_1|}{\varphi_1} - k_2 \dfrac{|\varphi_2|}{\varphi_2}}{1 - k_1 k_2 \dfrac{|\varphi_1|\varphi_2}{\varphi_1|\varphi_2|}} \right| .$$

If we set

$$\Lambda(z) = \frac{|\varphi_1|\varphi_2}{\varphi_1|\varphi_2|}$$

Then $|\Lambda(z)| \equiv 1$ and

[3] $$k_3 = \left|\frac{k_1 - k_2}{1 - k_1 k_2 \Lambda}\right| .$$

If $\Lambda(z) \not\equiv 1$ we obtain a contradiction as follows. Since $\gamma: z \mapsto (k_1 z - k_2)/(1 - k_1 k_2 z)$ is a real Möbius transformation, $\gamma\{|z| = 1\}$ has a unique minimum at $z = -1$, unless $\gamma(\{|z| = 1\}) = \{|z| = k_3\}$. In the latter case, $\gamma(-1) = -\gamma(1) = -k_3$ which implies that $k_1^2 = 1$ which is impossible.

Thus $\Lambda(z) \equiv 1$ and $\arg \varphi_1 = \arg \varphi_2 \pmod{2\pi}$ locally. We immediately obtain $\varphi_1 = \pm \varphi_2$ and S_3 lies on the (unique) Teichmüller line through S_1 and S_2. ///

The proof has the immediate consequence

Corollary: Every line in $T_{g,n,0}$ is a Teichmüller line.

(3.3) Hyperbolic transformations and invariant lines

We now propose to show that hyperbolic transformations may be characterized, among the elements of infinite order in Mod, by the property of having invariant lines. Precisely, we have

Theorem: *Let S_σ have non-excluded type $(g,n,0)$ and $f \in \mathrm{Diff}_+ S_\sigma$. If g_f has infinite order then the following are equivalent:*

(i.) *g_f is hyperbolic and $K[f_\sigma] = \overline{K}(f)$*

(ii.) *there exists a line L through S_σ so that $g_f(L) = L$.*

Proof: Since $Mod(g,n,0)$ acts discontinuously on $T_{g,n,0}$, g_f has no fixed point and S_σ, $g_f(S_\sigma)$ and $g_f^2(S_\sigma)$ are distinct.

(i $\Rightarrow$ ii) If $K[f_\sigma] = \overline{K}(f)$, let S_{σ_i} be the midpoint of the segment from $g_f^{i-1}(S_\sigma)$ to $g_f^i(S_\sigma)$. Then

$$d_T(S_\sigma, S_{\sigma_1}) = d_T(S_{\sigma_1}, g_f(S_\sigma)) = \frac{1}{2} d_T(S_\sigma, g_f(S_\sigma))$$

and

$$d_T(g_f(S_\sigma), S_{\sigma_2}) = d_T(S_{\sigma_2}, g_f^2(S_\sigma)) = \frac{1}{2} d_T(g_f(S_\sigma), g_f^2(S_\sigma)).$$

Because g_f is an isometry of $T_{g,n,0}$,

$$d_T(g_f(S_\sigma), S_{\sigma_2}) = d_T(S_{\sigma_2}, g_f^2(S_\sigma)) = \frac{1}{2} \log \overline{K}(f).$$

and consequently,

[4] $$g_f(S_{\sigma_1}) = S_{\sigma_2}.$$

The triangle inequality then gives

[5] $$d_T(S_{\sigma_1}, S_{\sigma_2}) \leq \log \overline{K}(f).$$

Since $\log \overline{K}(f) = \inf \{d_T(S, g_f(S)) \mid S \in T_{g,n,0}\}$, [4] and [5] show that

$$d_T(S_{\sigma_1}, S_{\sigma_2}) = \log \overline{K}(f)$$

and S_{σ_1}, S_{σ_2} and $g_f(S_\sigma)$ lie on a common line which must be L. S_{σ_1} and $g_f(S_\sigma)$ lie on L. L is therefore g_f-invariant.

(ii $\Rightarrow$ i) Since L is g_f-invariant, $g_f^k(S_\sigma) \in L$. Suppose $S_\tau \in T_{g,n,0}$ then by the triangle inequality and the additivity of

d_T along L

$$
\begin{aligned}
kd_T(S_\sigma, g_f(S_\sigma)) &= \sum_{i=1}^{k} d_T(g_f^{i-1}(S_\sigma), g_f^{i}(S_\sigma)) \\
&= d_T(S_\sigma, g_f^{k}(S_\sigma)) \\
&\leq d_T(S_\sigma, S_\tau) + d_T(S_\tau, g_f^{k}(S_\tau)) + d_T(g_f^{k}(S_\tau), g_f^{k}(S_\sigma)) \\
&\leq d_T(S_\sigma, S_\tau) + \sum_{i=1}^{k} d_T(g_f^{i-1}(S_\tau), g_f^{i}(S_\tau)) + d_T(g_f^{k}(S_\tau), g_f^{k}(S_\sigma)) \\
&= 2d_T(S_\sigma, S_\tau) + kd_T(S_\tau, g_f(S_\tau)).
\end{aligned}
$$

Thus $d_T(S_\sigma, g_f(S_\sigma)) \leq d_T(S_\tau, g_f(S_\tau))$ and $\overline{K}(f) = K[f_\sigma]$. Further, g_f is not elliptic, hence must be hyperbolic. ///

<u>Corollary</u>: If g_f is hyperbolic, then there exist infinitely many distinct conformal structures S_σ at which $\overline{K}(f)$ is attained.

<u>Proof</u>: It is attained at each point of L, but the points in $R(g,n,0)$ with conformal structures equivalent to a given surface S_σ is the $\mathrm{Mod}(g,n,0)$-orbit of S_σ. The orbit is countable, L is not. ///

(3.4) <u>Pseudo-Anosov diffeomorphisms</u>

An <u>Anosov diffeomorphism</u> $f: S \to S$ is a diffeomorphism so that for some pair $(\mathcal{F}_1, \mathcal{F}_2)$ of transverse measured foliations of S_τ and some $K > 1$, we have

(A-1) the image of a leaf of $\mathcal{F}_i$ is again a leaf of $\mathcal{F}_i$

(A-2) the distance δ between leaves ψ_1^i and ψ_2^i of $\mathcal{F}_i$ satisfy $\delta(f(\psi_1^2), f(\psi_2^2)) = K^{1/2}\delta(\psi_1^1, \psi_2^1)$ and

$$\delta(f(\psi_1^2), f(\psi_2^2)) = K^{-1/2}\delta(\psi_1^2, \psi_2^2) \quad .$$

A diffeomorphism f, with isolated singularities forming a set b_f, is <u>pseudo-Anosov</u> if:

(PA-1) $f(S\backslash b_f)$ is Anosov.

(PA-2) The $\mathcal{F}_i$ have identical singularities at $z \in b_f$ homeomorphic to the singularity of the horizontal foliation associated to a quadratic differential.

In this section we show that hyperbolic transformations in Mod are induced by pseudo-Anosov diffeomorphisms.

If $f: S_1 \to S$ is a Teichmüller mapping then $f = h \circ i$ where $i: S_1 \to (S_1, \frac{\omega}{\|\omega\|}, \|\omega\|)$ and h is conformal. If $\zeta = \xi + iy = \int \omega^{1/2}$ is the ω-coordinate on S then $\zeta' = K\xi + iy = \xi' + iy'$ with

$K = \dfrac{1 + ||\omega||}{1 - ||\omega||}$ is the local coordinate on $(S_1, \dfrac{\omega}{||\omega||}, ||\omega||)$

induced by the Teichmüller deformation. $\zeta_2 = \zeta' \circ h$ is the local coordinate on S_2 induced by the Teichmüller mapping f. We call $\omega/\|\omega\|$ the <u>initial</u> (and $d\zeta_2^2/\|d\zeta_2^2\|$ the <u>terminal</u>) <u>quadratic differential</u> for the Teichmüller mapping f.

<u>Theorem 1</u>: <u>Let</u> S_σ <u>be a Riemann surface of non-excluded type</u> $(g,n,0)$ <u>and</u> $f \in \mathrm{Diff}_+ S_\sigma$. <u>If</u> f <u>is not isotopic to a periodic self-map of</u> S_σ <u>then the following conditions are equivalent</u>:

(i) $\overline{K}(f) = K[f_\sigma]$

(ii) $f_\sigma^2 = f_\sigma \circ f_\sigma$ <u>is a Teichmüller mapping and</u>

$$K[f_\sigma^2] = K[f_\sigma]^2$$

and

(iii) <u>If ω and ω' are the initial and terminal quadratic differentials for</u> f_σ , <u>then</u> $\omega = \omega'$.

<u>Proof</u>: (i $\Rightarrow$ ii) Since f is not isotopic to a periodic self-map of S_σ , $g_f \in \mathrm{Mod}(g,n,0)$ is not elliptic by Corollary 2 to Theorem 2.1. It follows that $\overline{K}(f)$ - if attained - is non-zero. Thus if $K[f_\sigma] = \overline{K}(f)$, f_σ must be a Teichmüller mapping and g_f must be hyperbolic. $g_f(S_\sigma)$ is conformally equivalent to S_σ hence g_f may be applied again to $g_f(S_\sigma)$ to obtain

$$d_T(S_\sigma, g_f^2(S_\sigma)) \leq 2d_T(S_\sigma, g_f(S_\sigma))$$

or

$$K[f_\sigma^2] \leq K[f_\sigma]^2.$$

By Theorem (3.3), S_σ, $g_f(S_\sigma)$ and $g_f^2(S_\sigma)$ lie on a line in $T_{g,n,0}$, hence equality must hold in the previous two equations. It follows immediately that $f_\sigma \circ f_\sigma$ is a Teichmüller mapping.

(ii $\Rightarrow$ i) If $K[f_\sigma^2] = K[f_\sigma]^2$, then S_σ , $g_f(S_\sigma)$ and $g_f^2(S_\sigma)$ lie on a line L, which is therefore invariant under g_f. By Theorem (3.3), $\overline{K}(f)$ is attained at each point S_σ on L and g_f is hyperbolic.

To show that (iii) is equivalent to (ii) we first note that by the composition law for Beltrami coefficients (Equation 8 in §3.5 of Chapter I),

$$\mu_{f_\sigma^2}(z) = \frac{\mu_{f_\sigma}(z) + \mu_{f_\sigma}(f_\sigma(z))\exp(-2i\arg(f_\sigma)_z(z))}{1 + \overline{\mu}_{f_\sigma}(z)\mu_{f_\sigma}(f_\sigma(z))\exp(-2i\arg(f_\sigma)_z(z))}$$

in any holomorphic local coordinate z on S_σ. We simplify this expression as follows: set $\mu = \mu_{f_\sigma}(z)$ and

$$\hat{\mu}(z) = \mu_{f_\sigma}(f_\sigma(z))\exp(-2i\ \arg(f_\sigma)_z(z)).$$

Since f_σ is a Teichmüller mapping with astigmatism

$$K = K[f_\sigma] = \frac{1+k}{1-k},$$

$|\mu| = |\hat{\mu}| = k$ except at a finite number of points. We than have

$$\left|\mu_{f_\sigma^2}(z)\right|^2 = \left|\frac{\overline{\mu} + \hat{\mu}}{1 + \overline{\mu}\hat{\mu}}\right|^2$$

$$= \frac{2k^2 + 2\ \mathrm{Re}\ \overline{\mu}\hat{\mu}}{1 + k^4 + 2\ \mathrm{Re}\ \overline{\mu}\hat{\mu}}.$$

(ii $\Longrightarrow$ iii) If $K[f_\sigma^2] = K[f_\sigma]^2$ then

$$\left|\mu_{f_\sigma^2}\right|^2 = \frac{4k^2}{(k^2+1)^2}$$

which implies that $\mathrm{Re}\ \overline{\mu}\hat{\mu} = k^2$ or $\mu = \hat{\mu}$. But the composition law applied to $f_\sigma \circ (f_\sigma)^{-1}$ shows that $\hat{\mu} = -\mu_{f_\sigma^{-1}}$, the Beltrami coefficient associated to f_σ^{-1}. If $\omega = \varphi(z)dz^2$ is the initial quadratic differential of f_σ, then $\mu_{f_\sigma^{-1}} = -k|\varphi|/\varphi$ is the Beltrami coefficient of f_σ^{-1}. It follows immediately that ω is the terminal differential of f_σ.

(iii $\Longrightarrow$ i) If the initial differential ω of f_σ coincides with the terminal differential then, reversing the above logic, $\hat{\mu} = \mu$ and

$$\mu_{f_\sigma^2} = \frac{4k}{(1 + k^2)} \frac{|\varphi|}{\varphi}$$

i.e., f_σ^2 is a Teichmüller mapping. ///

An almost immediate consequence of the above Theorem is Thurston's classification, up to isotopy, of the diffeomorphisms of a topologically finite orientable surface. We need some preliminaries. We have seen that a Teichmüller deformation

$$i: S \to (S, \frac{\omega}{\|\omega\|}, \|\omega\|)$$

is locally given by $\zeta'(p) = i(\zeta) = K\xi + i\eta$ where

$$\zeta(p) = \xi + i\eta = \int_{p_0}^{p} \omega^{1/2}$$

for $p \notin b_\omega$ and $K = (1 + \|\omega\|)/(1 - \|\omega\|)$. The induced coordinate ζ is most useful to display the compactification of the Teichmüller space by measured foliations.

If $f: S_\sigma \to S_\sigma$ and f_σ is the Teichmüller map in the class of f, then f_σ must globally preserve area for any measure dA. Let $f_\sigma = h \circ f_1$ where h is conformal and f_1 is a Teichmüller deformation $i: S \to (S, \frac{\omega}{\|\omega\|}, \|\omega\|)$. Since f_1 expands the area element dA_ω by K, h must diminish area by K. If g_f is hyperbolic and ω_1 is the terminal differential of f_σ then $\omega_1 = \omega$ and the ω_1-coordinate ζ'' at $f_\sigma(\zeta)$ is precisely

$$\zeta''(f_\sigma(\zeta)) = K^{1/2}\xi + iK^{-1/2}\eta.$$

We then have

<u>Theorem 2</u>: (Thurston)

If S has non-excluded type (g,n,m) and $f \in \mathrm{Diff}_+ S$ is not isotopic to a periodic self-map then either:

(i) f *is isotopic to a reducible self-map, or*

(ii) f *is isotopic to a pseudo-Anosov diffeomorphism and* g_f *is hyperbolic.*

Further, (i) and (ii) do not occur simultaneously.

<u>Proof</u>: We assume f is irreducible and not isotopic to a periodic self-map. It follows that $\overline{K}(f)$ is attained at a conformal structure S_σ of type $(g,n+m,0)$ by a Teichmüller map f_σ. g_f is hyperbolic on $T_{g,n+m,0}$ and f is isotopic to f_σ. By Theorem 1, the initial and terminal differentials ω and ω' of f_σ are identical. If $K = K[f_\sigma]$ and $\omega = d\zeta^2$ with $\zeta = \xi + i\eta$ being the $\frac{\omega}{\|\omega\|}$-coordinate on S_σ, then $\zeta'' = K^{-1/2}\zeta' = K^{1/2}\xi + iK^{-1/2}\eta$ is a holomorphic local coordinate at $f_\sigma(\zeta)$. We notice that horizontal (respectively, vertical) lines are mapped into horizontal (respectively, vertical) lines since $\omega = (d\zeta'')^2$. f_σ expands distance between vertical lines by a factor of $K^{1/2}$ and contracts distance between horizontal lines by $K^{-1/2}$. Since f_σ has precisely the singularity structure of a Teichmüller map, f_σ is pseudo-Anosov.

To show that (i) and (ii) do not occur simultaneously note that if f is isotopic to a reducible map then $\overline{K}(f)$ is not attained. If f is pseudo-Anosov and $p \in S$ is not a singularity of the $\mathcal{F}_i$, we may map a neighborhood of p into the plane as follows: p lies on leaves F_i of $\mathcal{F}_i$. Choose leaves $G_{i,1}$ and $G_{i,2}$ at distance ϵ from F_i. The square bounded by $G_{1,1}$, $G_{1,2}$, $G_{2,1}$ and $G_{2,2}$

may be mapped into the square $|x| \leq \epsilon$, $|y| \leq \epsilon$ with the distance between leaves preserved (see Figure 3 below). These charts define a conformal structure on S. In this conformal structure, f is a Teichmüller mapping, as is f^2. By the previous theorem, g_f is hyperbolic and $\overline{K}(f)$ is attained. This contradiction completes the proof. ///

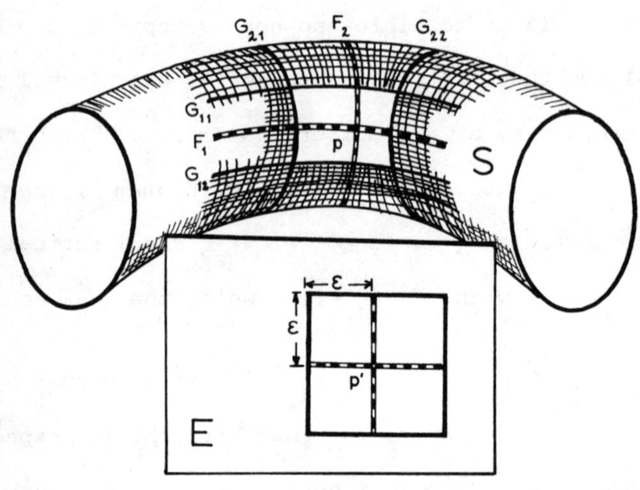

Figure 3

§4. Relative Teichmüller spaces and the fractional twist group

As we have noted several times previously, Royden [6] showed, in 1970, that the group of complex analytic automorphisms of the Teichmüller space $T_{g,0,0}$ is, for $g > 1$, a discrete group. In recent years, there has been renewed interest in the study of continuous group actions on the Teichmüller space, in particular those groups which contain the Teichmüller modular group as a discrete subgroup. Here we consider a specific group action, the fractional twist group and show that the action is transitive. An independent proof of the transitivity has been obtained by Wolpert [7] in the course of an extensive study of fractional twists. The proof given here uses less formidable machinery and produces less formidable results.

Let S_0 be a surface of finite non-excluded type (g,n,m) and $T_{g,n,m}$ be the Teichmüller space for that conformal type. Subordinate to any pants decomposition $\mathcal{P}$ of S_0 there is an associated Fenchel-Nielsen coordinate system for $T_{g,n,m}$ (c.f. Chapter II §3.2). These coordinates we have written as (L^{-1},Θ). The <u>Teichmüller space</u> $T(S_0,\partial S_0)$ <u>of</u> S_0 <u>relative to</u> ∂S_0 is the space of marked deformations of S_0 subject to the restriction that the intrinsic lengths of the border curves are held fixed. Since the lengths of all border curves are fixed, they may be omitted from the Fenchel-Nielsen coordinates of $S \in T(S_0,\partial S_0)$. The resulting coordinates (L_*^{-1},Θ) are independent of the choice of basepoint and lengths of the border curves, so we may write $T^*_{g,n,m}$ instead of $T(S_0,\partial S_0)$. Up to real analytic diffeomorphism, $T^*_{g,n,m} = T^*_{g,n+m,0} = T_{g,n+m,0}$.

(4.1) <u>Fractional twist maps</u>

Let α be a simple non-trivial oriented loop on S_0. We may choose a pants decomposition $\mathcal{P}$ of S_0 for which $\alpha = \alpha_i$ is a dissecting curve. In the Fenchel-Nielsen coordinates (L,Θ) on $T_{g,n,m}$, the Dehn twist about α_i is the map

$$\tau_\alpha : T_{g,n,m} \to T_{g,n,m}$$
$$: (L^{-1},\Theta) \mapsto (L^{-1},\Theta+2\pi e_i)$$

where e_i is the row vector with entries $(e_i)_j = \delta_{ij}$. The <u>fractional twist of angle</u> θ <u>about</u> α is the map

$$\tau_\alpha(\theta): (L^{-1},\Theta) \mapsto (L^{-1},\Theta + \theta e_i).$$

Twisting about a border curve gives a trivial action on $T_{g,n,m}$. Further, $\tau_\alpha(\theta)$ commutes with projection onto $T^*_{g,n,m}$.

A fractional twist map is generically a linear or horocyclic interpolation (or flow through) the orbit τ_α^n. We have chosen an interpolation which is linear in Fenchel-Nielsen coordinates although it is not the only natural interpolation (c.f. Marden and Masur [5]).

Vector space projections yield the following trivial but useful result.

<u>Theorem</u>: <u>Let</u> (g,n,m) <u>be a non-excluded type and</u> $\mathcal{P}$ <u>a pants decomposition of</u> $S \in T_{g,n,m}$ <u>with dissecting curves</u> $\alpha_1,\ldots,\alpha_k$. <u>There exists a trivial codimension</u> m <u>foliation</u> $\mathcal{F}$ <u>of</u> $T_{g,n,m}$ <u>so that</u>:

(i) <u>each leaf</u> $f \in \mathcal{F}$ <u>is a</u> $T(S_0,\partial S_0) = T^*_{g',n',m'} = T_{g',n'+m',0}$

(ii) f <u>is stable under the free abelian group</u> $\mathcal{T}_{\mathbb{R}}$ <u>generated by</u> $\{\tau_{\alpha_i}(\theta_i) \mid \theta_i \in \mathbb{R} \text{ and } i = 1,\ldots,k\}$

(iii) <u>the action of</u> $\mathcal{T}_R$ <u>commutes with projection onto</u> $T_{g,n+m,0}$

(iv) <u>the map</u>

$$\mathbb{R} \times T_{g,n,m} \to T_{g,n,m}$$

$$(\theta, S) \mapsto \tau_{\alpha_i}(\theta)(S)$$

is a non-singular real analytic flow.

The Dehn twists in $\mathcal{T}_{\mathbb{R}}$ form a subgroup $\mathcal{T}_{\mathbb{Z}} = \{g \in \mathcal{T}_R \mid \theta_i \in 2\pi Z\}$. $\mathcal{T}_{\mathbb{Z}}$ is clearly a lattice in the real algebraic group $\mathcal{T}_{\mathbb{R}}$.

(4.2) The asymptotic behavior of lengths of curves

$S \in T_{g,n,m}$ be of non-excluded type and α and β be simple non-homotopic loops on S. Recall that the geometric intersection number $\#(\alpha,\beta)$ is the infimum of the cardinality of $\alpha' \cap \beta'$ where α' (respectively β') ranges over the free homotopy class of α (respectively β). If S has either a complete hyperbolic metric or intrinsic metric, $\#(\alpha,\beta)$ is attained when α and β are geodesic in their free homotopy classes.

Assume α is not retractible into S or into a puncture and set $S_\theta = \tau_\alpha(\theta)(S)$.

Lemma 1: If β is homotopic to α or $\#(\beta,\alpha) = 0$, then $\ell[\beta,S_\theta] = \ell[\beta,S]$ and $\ell_I[\beta,S_\theta] = \ell_I[\beta,S]$. Here $\ell[\beta,S_1]$ (respectively $\ell_I[\beta,S_1]$) is the hyperbolic (respectively, intrinsic) length of the geodesic in the free homotopy class of β on the Riemann surface S_1.

Proof: Since $\ell_I[\beta,S] = \ell[\beta,S^d]$, we need only prove the lemma for hyperbolic length. If β is homotopic to α then $\ell[\alpha,S_\theta] = \ell[\alpha,S] = \ell[\beta,S] = \ell[\beta,S_\theta]$ as is obvious from the Fenchel-Nielsen coordinates with respect to any decomposition $\mathcal{P}$ for which

α is a dissecting curve.

If β is not homotopic to α, then β is either a border curve, surrounds a puncture or β and α are dissecting curves for some decomposition $\mathcal{P}$ of S. In any case, $\ell[\beta, S]$ is a slot function, for Fenchel-Nielsen coordinates on $T_{g,n,m}$, which is invariant under $\tau_\alpha(\theta)$. ///

Lemma 2: There is some $K > 1$ depending only on α and S, so that

$$K^{-1}\ell[\beta, S_{\theta'}] \leq \ell[\beta, S_\theta] \leq K\ell[\beta, S_{\theta'}]$$

where $\theta' = 2\pi[\theta/2\pi]$ and $[\ \ ]$ denotes integral part.

Proof: $\{S_\theta \mid 0 \leq \theta \leq 2\pi\}$ is a compact subset of $T_{g,n,m}$. Consequently the Teichmüller map $f_\theta : S \to S_\theta$ satisfies $K[f_\theta] \leq K < \infty$ for $0 \leq \theta \leq 2\pi$. By Wolpert's Theorem (Chapter II, §1.3 Theorem 4), $\ell[\beta, S_\theta] \leq K\ell[\beta, S]$. Since the map $\tau_\alpha(2\pi n)$ is a Teichmüller isometry (Chapter II, §2.1 Theorem 2), the Teichmüller map $h(\theta) : S_\theta \to S_{\theta'}$ satisfies $K[h(\theta)] = K[f(\theta - \theta')] \leq K$ and the result follows from Wolpert's Theorem since $K[h^{-1}(\theta)] = K[h(\theta)]$. ///

The two previous lemmas show that $\ell[\beta, S_\theta]$ is asymptotically determined by $\ell[\beta, S_{2\pi n}]$. To determine the latter quantity, we first need a lemma from hyperbolic geometry.

Lemma 3: Let $\pi : U \to U/G = S$ be the universal cover map. If L denotes the positive imaginary axis and $\pi(L)$ is a simple loop α on S then $d(L, \gamma_n(L)) \to \infty$ for any sequence $\gamma_n \in G$ for which the $\gamma_n(L)$ are distinct.

Proof: Assume $\zeta \in L$ and $\zeta_n \in \gamma_n(L)$ are chosen so that

$d(L,\gamma_n(L)) = d(\zeta,\zeta_n)$. On $\gamma_n(L)$ there is some ζ_n' which satisfies

(i) $d(\zeta_n,\zeta_n') < \ell[\alpha,S]$

(ii) $\zeta_n' = \eta_n(\zeta)$ for some $\eta_n \in G$.

By the triangle inequality,

$$d(\zeta,\zeta_n') < d(\zeta,\zeta_n) + \ell[\beta,S]$$

since the $\gamma_n(L)$ are distinct, so are the η_n. The discontinuity of G implies $d(\zeta,\zeta_n') \to \infty$, hence so does $d(\zeta,\zeta_n)$. ///

<u>Theorem</u>: If $\#(\beta,\alpha) \geq 1$, then

$$\lim_{\theta\to\pm\infty} \ell_I[\beta,\tau_\alpha(\theta)S] = \lim_{\theta\to\pm\infty} \ell[\beta,\tau_\alpha(\theta)S] = \infty .$$

<u>Proof</u>: Again, we need only prove the result for the hyperbolic metric and by Lemma 2, we need only show $\lim_{n\to\pm\infty} [\beta,S_{2\pi n}] = \infty$.

Let $\pi:U \to U/G = S$. We may assume that α lifts to the positive imaginary axis L. Since $\#(\beta,\alpha) \neq 0$, there is some lift $\tilde{\beta}$ of β which intersects L; we may assume $\tilde{\beta}$ is a hyperbolic line. $\tilde{\beta}$ also intersects some $\gamma_0(L) \neq L$. Let η be a primary deck transformation with axis $\tilde{\beta}$. As in Chapter II, Theorem (3.2), if $\gamma \in G$ is a primary transformation fixing L, then the twist $\tau_\alpha(2\pi n)$ may be realized by translating by γ^n the geometry on one side of L with respect to the other side. This translation is extended by group invariance to a translation along all images of L. In particular, the distinct translates $\gamma_k(L)$ are permuted by $\tau_\alpha(2\pi n)$ and $\ell[\beta,S_{2\pi n}] \to d(L,\gamma_k(L))$ for some choice of $\gamma_n \in G$

so that the $\gamma_n(L)$ are distinct. It follows from Lemma 3 that $\lim_{\theta\to\pm\infty} \ell[\beta, S_\theta] \to \infty$.

(4.3) <u>Transitivity of the fractional twist group</u>

Let S_o have non-excluded type (g,n,m) and fix some pants decomposition $\mathcal{P}$ of S_o with dissecting curves $\alpha_1,\ldots,\alpha_k$. Assume $S_i \in T_{g,n,m}$ have Fenchel-Nielsen coordinates (L_i^{-1}, Θ_i) with respect to $\mathcal{P}$. An immediate consequence of the definition of the fractional twist action is the following

<u>Lemma</u>: If $L_1^{-1} = L_2^{-1}$ and $(\varphi_1,\ldots,\varphi_k) = \Theta_2 - \Theta_1$ then $g = \prod_{i=1}^{k} \tau_\alpha(\varphi_i)$ commute.

There is no hope that the fractional twist group be transitive on $T_{g,n,m}$ if $m > 0$ since fractional twist maps cannot alter the lengths of border curves. Henceforth we will restrict our attention to the relative Teichmüller space $T^*_{g,n,m}$ using the coordinates (L_*^{-1}, Θ).

<u>Theorem</u>: <u>If</u> (g,n,m) <u>is a non-excluded type, then the fractional twist group is transitive on</u> $T^*_{g,n,m}$.

<u>Proof</u>: Let $\mathcal{P}$ be a pants decomposition of $S \in T_{g,n,m}$ and let $S_i \in T^*_{g,n,m}$ have Fenchel-Nielsen coordinates $((L_i)_*^{-1}, \Theta_i)$ where $(L_i)_*^{-1} = (\ell_{i,1}^{-1},\ldots,\ell_{i,k}^{-1})$. Define $S_1 \leq S_2$ if $\ell_{i,j} \leq \ell_{2,j}$ for $j \leq k$.

Each dissecting curve α_i lies in the border of one or two pants D_p as shown in Figure 4. Notice in Figure 4 that, in either case, we may choose a simple loop β_i so that $\#(\beta_i,\alpha_i) = 1$ or 2 and $\#(\beta_i,\alpha_j) = 0$ if $i \neq j$. It follows from Lemma 1 and the Theorem

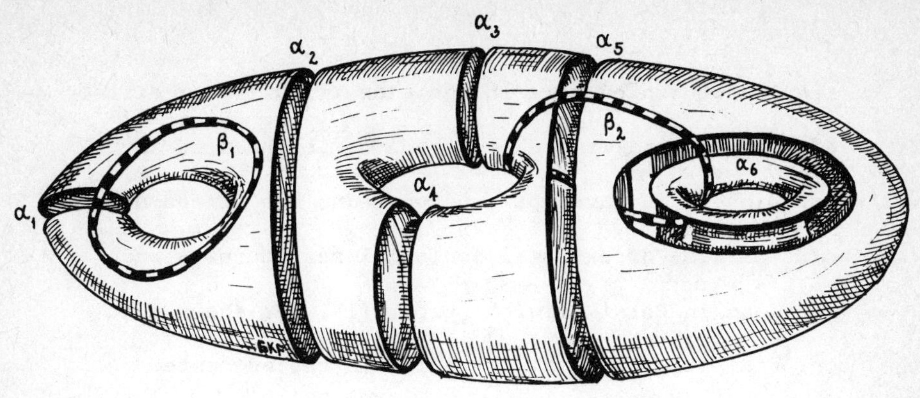

Figure 4

in 4.2 that $\ell[\alpha_j, \tau_{\beta_i}(\theta)S] = \ell[\alpha_j, S]$ if $i \neq j$ and $\ell[\alpha_i, \tau_{\beta_i}(\theta)S] \to \infty$ as $|\theta| \to \infty$. Since $\ell[\alpha_i, \tau_{\beta_i}(\theta)S]$ is a continuous function of θ, $\ell[\alpha_i, \tau_{\beta_i}(\theta)S]$ takes on all values $x \geq \ell[\alpha_i, S]$.

Choose $\lambda_i \geq \max\{\ell_{i,1}, \ell_{i,2}\}$. If $S = S_1$ then, for some choice of $\psi_1, \ldots, \psi_k \in \mathbb{R}$, the above argument shows that

$$\ell[\alpha_i, \tau_{\beta_k}(\psi_k) \cdots \cdot \tau_{\beta_2}(\psi_2)\tau_{\beta_1}(\psi_1)S_1] = \lambda_i \quad \text{for} \quad i = 1, \ldots, k .$$

Let $S_1' = [\prod_{i=1}^{k} \tau_{\beta_i}(\psi_i)]S_1$. Symmetrically there exist $\psi_i' \in \mathbb{R}$ so that $S_2' = [\prod_{i=1}^{k} \tau_{\beta_i}(\psi_i')]S_2$ satisfies $\ell[\alpha_i, S_2'] = \lambda_i$. By the previous Lemma, $S_1' = gS_2'$ for some product g of fractional twists. It follows that the orbit of S_1 under the fractional twist group contains S_2 and the fractional twist group is transitive on $T^*_{g,n,m}$. ///

§5 Concluding Remarks

The results presented here on the classification of diffeomorphisms of surfaces and/or modular transformations are, in a sense, not complete. We have not studied reducible diffeomorphisms from the point of view of extremal conformal structures. Such a study may be found in Bers' original paper [1]. He shows that extremal conformal structures always exist in the augmented Teichmüller space. In fact, one may show that any minimizing sequence for Bers' extremal problem clusters, in T(S), to an extremal conformal structure. The converse is also true.

We have seen that a hyperbolic modular transformation g has an invariant line along which $d_T(S,g(S))$ is minimized. Earle [2] has shown that every critical point of the map

$$d_g : T(S) \to \mathbb{R}$$
$$: S \mapsto d_T(S,g(S))$$

is a local minimum whenever g is hyperbolic. Although many mathematicians have asserted that the invariant line is unique, no proof has ever been seen by this author.

Our discussion of fractional twists is a very elementary introduction to a fascinating new method due to Thurston for altering the conformal structure of a surface. Instead of doing fractional twists along simple closed geodesics, he fractional twists along simple open geodesics - the operation is called earthquaking along a lamination. Kerchkoff has recently announced a proof of the Nielsen realization problem using the Thurston technique. Nielsen's problem, stated in the context of this

Chapter, is: suppose $\Gamma < M(g,0,0)$ is finite, does there exist $S \in T_{g,0,0}$ so that every element of Γ fixes S. It is worthwhile to compare Nielsen's problem with the results in §2.

$T^*_{g,n,m}$ may be given a complex analytic structure using the Fenchel-Nielsen coordinates in the form $z_j = \theta_j + i\ell_j^{-1}$. This structure is most decidedly not the usual complex analytic structure, in fact $T^*_{g,n,m}$ with this structure is biholomorphic to a polydisc. It is unknown whether this structure depends on the choice of pants decomposition or whether Mod acts holomorphically.

The closure of the fractional twist group in the homeomorphism group of $T^*_{g,n,m}$ is an important, but unstudied, group. A major question is whether it is locally compact.

References

[1] L. Bers, An extremal problem for quasiconformal mappings and a theorem by Thurston, Acta Math., 141(1978), 73-98.

[2] C. Earle, to appear.

[3] S. Kravetz, On the geometry of Teichmüller spaces and the structure of their modular groups, Ann. Acad. Sci. Fenn., 278(1959), 1-35.

[4] W. Harvey, Geometric structure of surface mapping-class groups, to appear.

[5] A. Marden and H. Masur, A foliation of Teichmüller space by twist invariant discs, Math. Scand., 36(1975), 221-228.

[6] H. Royden, Automorphisms and isometries of Teichmüller space, Ann. of Math. Studies, 66(1971), 369-383.

[7] S. Wolpert, The Fenchel-Nielsen twist deformation, to appear.

INDEX

Vol. 670: Fonctions de Plusieurs Variables Complexes III, Proceedings, 1977. Edité par F. Norguet. XII, 394 pages. 1978.

Vol. 671: R. T. Smythe and J. C. Wierman, First-Passage Perculation on the Square Lattice. VIII, 196 pages. 1978.

Vol. 672: R. L. Taylor, Stochastic Convergence of Weighted Sums of Random Elements in Linear Spaces. VII, 216 pages. 1978.

Vol. 673: Algebraic Topology, Proceedings 1977. Edited by P. Hoffman, R. Piccinini and D. Sjerve. VI, 278 pages. 1978.

Vol. 674: Z. Fiedorowicz and S. Priddy, Homology of Classical Groups Over Finite Fields and Their Associated Infinite Loop Spaces. VI, 434 pages. 1978.

Vol. 675: J. Galambos and S. Kotz, Characterizations of Probability Distributions. VIII, 169 pages. 1978.

Vol. 676: Differential Geometrical Methods in Mathematical Physics II, Proceedings, 1977. Edited by K. Bleuler, H. R. Petry and A. Reetz. VI, 626 pages. 1978.

Vol. 677: Séminaire Bourbaki, vol. 1976/77, Exposés 489–506. IV, 264 pages. 1978.

Vol. 678: D. Dacunha-Castelle, H. Heyer et B. Roynette. Ecole d'Eté de Probabilités de Saint-Flour. VII-1977. Edité par P. L. Hennequin. IX, 379 pages. 1978.

Vol. 679: Numerical Treatment of Differential Equations in Applications, Proceedings, 1977. Edited by R. Ansorge and W. Törnig. IX, 163 pages. 1978.

Vol. 680: Mathematical Control Theory, Proceedings, 1977. Edited by W. A. Coppel. IX, 257 pages. 1978.

Vol. 681: Séminaire de Théorie du Potentiel Paris, No. 3, Directeurs: M. Brelot, G. Choquet et J. Deny. Rédacteurs: F. Hirsch et G. Mokobodzki. VII, 294 pages. 1978.

Vol. 682: G. D. James, The Representation Theory of the Symmetric Groups. V, 156 pages. 1978.

Vol. 683: Variétés Analytiques Compactes, Proceedings, 1977. Edité par Y. Hervier et A. Hirschowitz. V, 248 pages. 1978.

Vol. 684: E. E. Rosinger, Distributions and Nonlinear Partial Differential Equations. XI, 146 pages. 1978.

Vol. 685: Knot Theory, Proceedings, 1977. Edited by J. C. Hausmann. VII, 311 pages. 1978.

Vol. 686: Combinatorial Mathematics, Proceedings, 1977. Edited by D. A. Holton and J. Seberry. IX, 353 pages. 1978.

Vol. 687: Algebraic Geometry, Proceedings, 1977. Edited by L. D. Olson. V, 244 pages. 1978.

Vol. 688: J. Dydak and J. Segal, Shape Theory. VI, 150 pages. 1978.

Vol. 689: Cabal Seminar 76–77, Proceedings, 1976–77. Edited by A.S. Kechris and Y. N. Moschovakis. V, 282 pages. 1978.

Vol. 690: W. J. J. Rey, Robust Statistical Methods. VI, 128 pages. 1978.

Vol. 691: G. Viennot, Algèbres de Lie Libres et Monoïdes Libres. III, 124 pages. 1978.

Vol. 692: T. Husain and S. M. Khaleelulla, Barrelledness in Topological and Ordered Vector Spaces. IX, 258 pages. 1978.

Vol. 693: Hilbert Space Operators, Proceedings, 1977. Edited by J. M. Bachar Jr. and D. W. Hadwin. VIII, 184 pages. 1978.

Vol. 694: Séminaire Pierre Lelong – Henri Skoda (Analyse) Année 1976/77. VII, 334 pages. 1978.

Vol. 695: Measure Theory Applications to Stochastic Analysis, Proceedings, 1977. Edited by G. Kallianpur and D. Kölzow. XII, 261 pages. 1978.

Vol. 696: P. J. Feinsilver, Special Functions, Probability Semigroups, and Hamiltonian Flows. VI, 112 pages. 1978.

Vol. 697: Topics in Algebra, Proceedings, 1978. Edited by M. F. Newman. XI, 229 pages. 1978.

Vol. 698: E. Grosswald, Bessel Polynomials. XIV, 182 pages. 1978.

Vol. 699: R. E. Greene and H.-H. Wu, Function Theory on Manifolds Which Possess a Pole. III, 215 pages. 1979.

Vol. 700: Module Theory, Proceedings, 1977. Edited by C. Faith and S. Wiegand. X, 239 pages. 1979.

Vol. 701: Functional Analysis Methods in Numerical Analysis, Proceedings, 1977. Edited by M. Zuhair Nashed. VII, 333 pages. 1979.

Vol. 702: Yuri N. Bibikov, Local Theory of Nonlinear Analytic Ordinary Differential Equations. IX, 147 pages. 1979.

Vol. 703: Equadiff IV, Proceedings, 1977. Edited by J. Fábera. XIX, 441 pages. 1979.

Vol. 704: Computing Methods in Applied Sciences and Engineering, 1977, I. Proceedings, 1977. Edited by R. Glowinski and J. L. Lions. VI, 391 pages. 1979.

Vol. 705: O. Forster und K. Knorr, Konstruktion verseller Familien kompakter komplexer Räume. VII, 141 Seiten. 1979.

Vol. 706: Probability Measures on Groups, Proceedings, 1978. Edited by H. Heyer. XIII, 348 pages. 1979.

Vol. 707: R. Zielke, Discontinuous Čebyšev Systems. VI, 111 pages. 1979.

Vol. 708: J. P. Jouanolou, Equations de Pfaff algébriques. V, 255 pages. 1979.

Vol. 709: Probability in Banach Spaces II. Proceedings, 1978. Edited by A. Beck. V, 205 pages. 1979.

Vol. 710: Séminaire Bourbaki vol. 1977/78, Exposés 507–524. IV, 328 pages. 1979.

Vol. 711: Asymptotic Analysis. Edited by F. Verhulst. V, 240 pages. 1979.

Vol. 712: Equations Différentielles et Systèmes de Pfaff dans le Champ Complexe. Edité par R. Gérard et J.-P. Ramis. V, 364 pages. 1979.

Vol. 713: Séminaire de Théorie du Potentiel, Paris No. 4. Edité par F. Hirsch et G. Mokobodzki. VII, 281 pages. 1979.

Vol. 714: J. Jacod, Calcul Stochastique et Problèmes de Martingales. X, 539 pages. 1979.

Vol. 715: Inder Bir S. Passi, Group Rings and Their Augmentation Ideals. VI, 137 pages. 1979.

Vol. 716: M. A. Scheunert, The Theory of Lie Superalgebras. X, 271 pages. 1979.

Vol. 717: Grosser, Bidualräume und Vervollständigungen von Banachmoduln. III, 209 pages. 1979.

Vol. 718: J. Ferrante and C. W. Rackoff, The Computational Complexity of Logical Theories. X, 243 pages. 1979.

Vol. 719: Categorial Topology, Proceedings, 1978. Edited by H. Herrlich and G. Preuß. XII, 420 pages. 1979.

Vol. 720: E. Dubinsky, The Structure of Nuclear Fréchet Spaces. V, 187 pages. 1979.

Vol. 721: Séminaire de Probabilités XIII. Proceedings, Strasbourg, 1977/78. Edité par C. Dellacherie, P. A. Meyer et M. Weil. VII, 647 pages. 1979.

Vol. 722: Topology of Low-Dimensional Manifolds. Proceedings, 1977. Edited by R. Fenn. VI, 154 pages. 1979.

Vol. 723: W. Brandal, Commutative Rings whose Finitely Generated Modules Decompose. II, 116 pages. 1979.

Vol. 724: D. Griffeath, Additive and Cancellative Interacting Particle Systems. V, 108 pages. 1979.

Vol. 725: Algèbres d'Opérateurs. Proceedings, 1978. Edité par P. de la Harpe. VII, 309 pages. 1979.

Vol. 726: Y.-C. Wong, Schwartz Spaces, Nuclear Spaces and Tensor Products. VI, 418 pages. 1979.

Vol. 727: Y. Saito, Spectral Representations for Schrödinger Operators With Long-Range Potentials. V, 149 pages. 1979.

Vol. 728: Non-Commutative Harmonic Analysis. Proceedings, 1978. Edited by J. Carmona and M. Vergne. V, 244 pages. 1979.

Vol. 729: Ergodic Theory. Proceedir and K. Jacobs. XII, 209 pages. 1979.

Vol. 730: Functional Differential Equations and Approximation of Fixed Points. Proceedings, 1978. Edited by H.-O. Peitgen and H.-O. Walther. XV, 503 pages. 1979.

Vol. 731: Y. Nakagami and M. Takesaki, Duality for Crossed Products of von Neumann Algebras. IX, 139 pages. 1979.

Vol. 732: Algebraic Geometry. Proceedings, 1978. Edited by K. Lønsted. IV, 658 pages. 1979.

Vol. 733: F. Bloom, Modern Differential Geometric Techniques in the Theory of Continuous Distributions of Dislocations. XII, 206 pages. 1979.

Vol. 734: Ring Theory, Waterloo, 1978. Proceedings, 1978. Edited by D. Handelman and J. Lawrence. XI, 352 pages. 1979.

Vol. 735: B. Aupetit, Propriétés Spectrales des Algèbres de Banach. XII, 192 pages. 1979.

Vol. 736: E. Behrends, M-Structure and the Banach-Stone Theorem. X, 217 pages. 1979.

Vol. 737: Volterra Equations. Proceedings 1978. Edited by S.-O. Londen and O. J. Staffans. VIII, 314 pages. 1979.

Vol. 738: P. E. Conner, Differentiable Periodic Maps. 2nd edition, IV, 181 pages. 1979.

Vol. 739: Analyse Harmonique sur les Groupes de Lie II. Proceedings, 1976–78. Edited by P. Eymard et al. VI, 646 pages. 1979.

Vol. 740: Séminaire d'Algèbre Paul Dubreil. Proceedings, 1977–78. Edited by M.-P. Malliavin. V, 456 pages. 1979.

Vol. 741: Algebraic Topology, Waterloo 1978. Proceedings. Edited by P. Hoffman and V. Snaith. XI, 655 pages. 1979.

Vol. 742: K. Clancey, Seminormal Operators. VII, 125 pages. 1979.

Vol. 743: Romanian-Finnish Seminar on Complex Analysis. Proceedings, 1976. Edited by C. Andreian Cazacu et al. XVI, 713 pages. 1979.

Vol. 744: I. Reiner and K. W. Roggenkamp, Integral Representations. VIII, 275 pages. 1979.

Vol. 745: D. K. Haley, Equational Compactness in Rings. III, 167 pages. 1979.

Vol. 746: P. Hoffman, τ-Rings and Wreath Product Representations. V, 148 pages. 1979.

Vol. 747: Complex Analysis, Joensuu 1978. Proceedings, 1978. Edited by I. Laine, O. Lehto and T. Sorvali. XV, 450 pages. 1979.

Vol. 748: Combinatorial Mathematics VI. Proceedings, 1978. Edited by A. F. Horadam and W. D. Wallis. IX, 206 pages. 1979.

Vol. 749: V. Girault and P.-A. Raviart, Finite Element Approximation of the Navier-Stokes Equations. VII, 200 pages. 1979.

Vol. 750: J. C. Jantzen, Moduln mit einem höchsten Gewicht. III, 195 Seiten. 1979.

Vol. 751: Number Theory, Carbondale 1979. Proceedings. Edited by M. B. Nathanson. V, 342 pages. 1979.

Vol. 752: M. Barr, *-Autonomous Categories. VI, 140 pages. 1979.

Vol. 753: Applications of Sheaves. Proceedings, 1977. Edited by M. Fourman, C. Mulvey and D. Scott. XIV, 779 pages. 1979.

Vol. 754: O. A. Laudal, Formal Moduli of Algebraic Structures. III, 161 pages. 1979.

Vol. 755: Global Analysis. Proceedings, 1978. Edited by M. Grmela and J. E. Marsden. VII, 377 pages. 1979.

Vol. 756: H. O. Cordes, Elliptic Pseudo-Differential Operators – An Abstract Theory. IX, 331 pages. 1979.

Vol. 757: Smoothing Techniques for Curve Estimation. Proceedings, 1979. Edited by Th. Gasser and M. Rosenblatt. V, 245 pages. 1979.

Vol. 758: C. Năstăsescu and F. Van Oystaeyen; Graded and Filtered Rings and Modules. X, 148 pages. 1979.

Degrees of Unsolvability: Structure and Theory. XIV, 216 pages. 1979.

Vol. 760: H.-O. Georgii, Canonical Gibbs Measures. VIII, 190 pages. 1979.

Vol. 761: K. Johannson, Homotopy Equivalences of 3-Manifolds with Boundaries. 2, 303 pages. 1979.

Vol. 762: D. H. Sattinger, Group Theoretic Methods in Bifurcation Theory. V, 241 pages. 1979.

Vol. 763: Algebraic Topology, Aarhus 1978. Proceedings, 1978. Edited by J. L. Dupont and H. Madsen. VI, 695 pages. 1979.

Vol. 764: B. Srinivasan, Representations of Finite Chevalley Groups. XI, 177 pages. 1979.

Vol. 765: Padé Approximation and its Applications. Proceedings, 1979. Edited by L. Wuytack. VI, 392 pages. 1979.

Vol. 766: T. tom Dieck, Transformation Groups and Representation Theory. VIII, 309 pages. 1979.

Vol. 767: M. Namba, Families of Meromorphic Functions on Compact Riemann Surfaces. XII, 284 pages. 1979.

Vol. 768: R. S. Doran and J. Wichmann, Approximate Identities and Factorization in Banach Modules. X, 305 pages. 1979.

Vol. 769: J. Flum, M. Ziegler, Topological Model Theory. X, 151 pages. 1980.

Vol. 770: Séminaire Bourbaki vol. 1978/79 Exposés 525–542. IV, 341 pages. 1980.

Vol. 771: Approximation Methods for Navier-Stokes Problems. Proceedings, 1979. Edited by R. Rautmann. XVI, 581 pages. 1980.

Vol. 772: J. P. Levine, Algebraic Structure of Knot Modules. XI, 104 pages. 1980.

Vol. 773: Numerical Analysis. Proceedings, 1979. Edited by G. A. Watson. X, 184 pages. 1980.

Vol. 774: R. Azencott, Y. Guivarc'h, R. F. Gundy, Ecole d'Eté de Probabilités de Saint-Flour VIII-1978. Edited by P. L. Hennequin. XIII, 334 pages. 1980.

Vol. 775: Geometric Methods in Mathematical Physics. Proceedings, 1979. Edited by G. Kaiser and J. E. Marsden. VII, 257 pages. 1980.

Vol. 776: B. Gross, Arithmetic on Elliptic Curves with Complex Multiplication. V, 95 pages. 1980.

Vol. 777: Séminaire sur les Singularités des Surfaces. Proceedings, 1976-1977. Edited by M. Demazure, H. Pinkham and B. Teissier. IX, 339 pages. 1980.

Vol. 778: SK_1 von Schiefkörpern. Proceedings, 1976. Edited by P. Draxl and M. Kneser. II, 124 pages. 1980.

Vol. 779: Euclidean Harmonic Analysis. Proceedings, 1979. Edited by J. J. Benedetto. III, 177 pages. 1980.

Vol. 780: L. Schwartz, Semi-Martingales sur des Variétés, et Martingales Conformes sur des Variétés Analytiques Complexes. XV, 132 pages. 1980.

Vol. 781: Harmonic Analysis Iraklion 1978. Proceedings 1978. Edited by N. Petridis, S. K. Pichorides and N. Varopoulos. V, 213 pages. 1980.

Vol. 782: Bifurcation and Nonlinear Eigenvalue Problems. Proceedings, 1978. Edited by C. Bardos, J. M. Lasry and M. Schatzman. VIII, 296 pages. 1980.

Vol. 783: A. Dinghas, Wertverteilung meromorpher Funktionen in ein- und mehrfach zusammenhängenden Gebieten. Edited by R. Nevanlinna and C. Andreian Cazacu. XIII, 145 pages. 1980.

Vol. 784: Séminaire de Probabilités XIV. Proceedings, 1978/79. Edited by J. Azéma and M. Yor. VIII, 546 pages. 1980.

Vol. 785: W. M. Schmidt, Diophantine Approximation. X, 299 pages. 1980.

Vol. 786: I. J. Maddox, Infinite Matrices of Operators. V, 122 pages. 1980.